This series publishes advanced monographs giving well-written presentations of the "state-of-the-art" in fields of mathematical research that have acquired the maturity needed for such a treatment. They are sufficiently self-contained to be accessible to more than just the intimate specialists of the subject, and sufficiently comprehensive to remain valuable references for many years. Besides the current state of knowledge in its field, an SMM volume should also describe its relevance to and interaction with neighbouring fields of mathematics, and give pointers to future directions of research.

More information about this series at http://www.springer.com/series/3733

Takeo Ohsawa

L^2 Approaches in Several Complex Variables

Development of Oka–Cartan Theory
by L^2 Estimates for the $\bar{\partial}$ Operator

 Springer

Takeo Ohsawa
Graduate School of Mathematics
Nagoya University
Nagoya, Japan

ISSN 1439-7382   ISSN 2196-9922 (electronic)
Springer Monographs in Mathematics
ISBN 978-4-431-56296-2   ISBN 978-4-431-55747-0 (eBook)
DOI 10.1007/978-4-431-55747-0

Springer Tokyo Heidelberg New York Dordrecht London

Printed on acid-free paper

Springer Japan KK is part of Springer Science+Business Media (www.springer.com)

Preface

As in the study of complex analysis of one variable, the general theory of several complex variables has manifold aspects. First, it provides a firm ground for systematic studies of special functions such as elliptic functions, theta functions, and modular functions. The general theory plays a role in confirming the existence and uniqueness of functions with prescribed zeros and poles. Another aspect is to give an insight into the connection between two different fields of mathematics by understanding how the tools work. The theory of sheaves bridged analysis and topology in such a way. In the construction of this basic theory of several complex variables, a particularly important contribution was made by two mathematicians, Kiyoshi Oka (1901–1978) and Henri Cartan (1904–2008). The theory of Oka and Cartan is condensed in a statement that the first cohomology of coherent analytic sheaves over $\mathbb{C}^n$ is zero. On the other hand, the method of PDE (partial differential equations) had turned out to be essential in the existence of conformal mappings. By this approach, the function theory on Riemann surfaces as one-dimensional complex manifolds was explored by H. Weyl. Weyl's method was developed on manifolds of higher dimension by K. Kodaira who generalized Riemann's condition for Abelian varieties by establishing a differential geometric characterization of nonsingular projective algebraic varieties. This PDE method, based on the L^2 estimates for the $\bar{\partial}$–operator, was generalized by J. Kohn, L. Hörmander, A. Andreotti, and E. Vesentini. As a result, it enabled us to see the results of Oka and Cartan in a much higher resolution. In particular, based on such a refinement, existence theorems for holomorphic functions with L^2 growth conditions have been obtained by Hörmander, H. Skoda, and others. The purpose of the present monograph is to report on some of the recent results in several complex variables obtained by the L^2 method which can be regarded as a continuation of these works. Among various topics including complex geometry, the Bergman kernel, and holomorphic foliations, a special emphasis is put on the extension theorems and its applications. In this topic, highlighted are the recent developments after the solution of a long-standing open question of N. Suita. It is an inequality between the Bergman kernel and the logarithmic capacity on Riemann surfaces, which was first proved by Z. Błocki for plane domains. Q. Guan and X.-Y. Zhou proved generalized variants and characterized those surfaces on which the inequality is strict. Their work gave the author a decisive impetus to start writing a survey to cover these remarkable achievements. As a result, he found an alternate proof of the inequality, based on hyperbolic geometry, which is presented in Chap. 3. However, the readers are recommended to have a glance at Chap. 4 first, where the questions on the Bergman kernels are described more systematically. (The author started to write the monograph from Chap. 4.) Since there have been a lot of subsequent progress concerning the materials in Chaps. 3 and 4 during the preparation of the manuscript, it became soon beyond the author's ability to give a satisfactory account of the whole development. So he will be happy to have a chance in the future to revise and enlarge this rather brief monograph.

Nagoya, Japan   Takeo Ohsawa
March 2015

Contents

Chapter 1
Basic Notions and Classical Results

Abstract As a preliminary, basic properties of holomorphic functions and complex manifolds are recalled. Beginning with the definitions and characterizations of holomorphic functions, we shall give an overview of the classical theorems in several complex variables, restricting ourselves to extremely important ones for the discussion in later chapters. Most of the materials presented here are contained in well-written textbooks such as Gunning and Rossi (Analytic functions of several complex variables. Prentice-Hall, Englewood Cliffs, xiv+317 pp, 1965), Hörmander (An introduction to complex analysis in several variables. North-Holland Mathematical Library, vol 7, 3rd edn. North-Holland, Amsterdam, xii+254 pp, 1990), Wells and Raymond (Differential analysis on complex manifolds, Third edition. With a new appendix by Oscar Garcia-Prada. Graduate Texts in Mathematics, vol 65. Springer, New York, 2008), Grauert and Remmert (Theory of Stein spaces, Translated from the German by Alan Huckleberry. Reprint of the 1979 translation. Classics in mathematics. Springer, Berlin, xxii+255 pp, 2004), Grauert and Remmert (Coherent analytic sheaves, Grundlehren der Mathematischen Wissenschaften, vol 265. Springer, Berlin, xviii+249 pp, 1984) and Noguchi (Analytic function theory of several variables —Elements of Oka's coherence, preprint (translated from Japanese)) (see also Demailly, Analytic methods in algebraic geometry. Surveys of modern mathematics, vol 1. International Press, Somerville; Higher Education Press, Beijing, viii+231 pp, 2012 and Ohsawa, Analysis of several complex variables. Translated from the Japanese by Shu Gilbert Nakamura. Translations of mathematical monographs. Iwanami series in modern mathematics, vol 211. American Mathematical Society, Providence, xviii+121 pp, 2002), so that only sketchy accounts are given for most of the proofs and historical backgrounds. An exception is Serre's duality theorem. It will be presented after an article of Laurent-Thiébaut and Leiterer (Nagoya Math J 154:141–156, 1999), since none of the above books contains its proof in full generality.

© Springer Japan 2015

T. Ohsawa, L^2 *Approaches in Several Complex Variables*, Springer Monographs in Mathematics, DOI 10.1007/978-4-431-55747-0_1

1

1.1 Functions and Domains over $\mathbb{C}^n$

1.1.1 Holomorphic Functions and Cauchy's Formula

Let n be a positive integer and let $\mathbb{C}\{z\}$ be the convergent power series ring in $z = (z_1, \ldots, z_n)$ with coefficients in $\mathbb{C}$. Since the nineteenth century, $\mathbb{C}\{z\}$ has been identified with the set of germs at $z = (0, \ldots, 0)$ of functions of a distinguished class in complex variables $z_1 = x_1 + iy_1, \ldots, z_n = x_n + iy_n$, i.e. the class of holomorphic functions. Recall that a function f on an open subset U of $\mathbb{C}^n$ is called a **holomorphic function** if the values of f are equal to those of a convergent power series in $z - a$ around each point a of U. The set of holomorphic functions on U will be denoted by $\mathcal{O}(U)$ and the germ of $f \in \mathcal{O}(U)$ at $a \in U$ by f_a. The most important formula for holomorphic functions is **Cauchy's formula**,

$$f(z) = \frac{1}{2\pi i} \int_{\partial D} \frac{f(\zeta)}{\zeta - z} d\zeta. \tag{1.1}$$

Here D is a bounded domain in $\mathbb{C}$ with C^1-smooth boundary, i.e., the boundary ∂D of D is the disjoint union of finitely many C^1-smooth closed curves, f is holomorphic on a neighborhood of the closure of D, $z \in D$ and the orientation of ∂D as a path of the integral is defined to be the direction which sees the interior of D on the left–hand side. Let $D_1, \ldots, D_n$ be bounded domains in $\mathbb{C}$ with C^1-smooth boundary. If $f(z) = f(z_1, \ldots, z_n)$ is a holomorphic function on $U \subset \mathbb{C}^n$, $U \supset \overline{D_1 \times \cdots \times D_n}$ and $z_j \in D_j$, then (1.1) is generalized to

$$f(z) = \left(\frac{1}{2\pi i}\right)^n \prod_{j=1}^{n} \left(\int_{\partial D_j} \frac{d\zeta_j}{\zeta_j - z_j} \right) f(\zeta_1, \ldots, \zeta_n). \tag{1.2}$$

The right–hand side of (1.2), say $\tilde{f}(z)$, is holomorphic on $\mathbb{C}^n \setminus (\cup_{j=1}^{n} \mathbb{C} \times \cdots \times \partial D_j \times \cdots \times \mathbb{C})$ even if f is only defined on $\partial D_1 \times \cdots \times \partial D_n$ and continuous there. Hence, if further f is continuously extended to a subset B of $\overline{D_1 \times \cdots \times D_n}$ with $\partial D_1 \times \cdots \times \partial D_n \subset B$ in such a way that (1.2) holds for all $z \in B^\circ$, then $\tilde{f}|_{D_1 \times \cdots \times D_n}$ is a holomorphic extension of $f|_{B^\circ}$. Here B° denotes the set of interior points of B. In particular, letting D_j be the unit disc $\mathbb{D} = \{\zeta \in \mathbb{C}; |\zeta| < 1\}$ and choosing B in such a way that

$$B^\circ = T_{R_1, R_2} := \{z \in \mathbb{D}^n; \max\{|z_1|, \max_{2 \le j \le n} R_1|z_j|\} < 1 \ or \ R_2|z_1| > 1\}$$

for $R_1, R_2 > 1$, one has:

Theorem 1.1 (Hartogs's continuation theorem). *If $n \geq 2$, the natural restriction map*

$$\mathscr{O}(\mathbb{D}^n) \rightarrow \mathscr{O}(T_{R_1, R_2})$$

is surjective.

Thus, Cauchy's integral formula is useful to solve the boundary value problem for holomorphic functions. A remarkable point is that the boundary values have to be given only along a special subset of the topological boundary. In the case of one complex variable, (1.1) is also useful to solve the boundary value problem of this type for harmonic functions, the Dirichlet problem, but only in special cases (e.g., Poisson's formula). The class of subharmonic functions is useful to solve it in full generality. We recall that a **subharmonic function** on a domain $D \subset \mathbb{C}$ is by definition an upper semicontinuous function $u : D \rightarrow [-\infty, \infty)$ such that, for any disc $\mathbb{D}(c, r) := \{z \in \mathbb{C}; |z - c| < r\}$ in D, and for any harmonic function h on a neighborhood of $\overline{\mathbb{D}}(c, r)$ satisfying $u(z) \leq h(z)$ on $\partial \mathbb{D}(c, r)$, $u(z) \leq h(z)$ holds on $\mathbb{D}(c, r)$. We recall also that h is harmonic if and only if h is locally the real part of a holomorphic function (in the case of one variable). A standard method for finding a harmonic function with a given boundary value is to take the supremum of the family of subharmonic functions whose boundary values are inferior to the given function, and this method can be naturally extended to solve higher–dimensional Dirichlet problems.

Subharmonic functions also arise naturally as $\log |f|$ for any holomorphic function f. An observation closely related to this and the discovery of Theorem 1.1 is that, given any element

$$\sigma = \sum_{j,k=0}^{\infty} a_{jk} z_1^j z_2^k \in \mathbb{C}\{z_1, z_2\},$$

the lower envelope $\tilde{r}(z_2)$ of the radii of convergence $r(z_2)$ of the series

$$\sum_j \left(\sum_{k=0}^{\infty} a_{jk} z_2^k \right) z_1^j$$

in z_1 ($\tilde{r}(c) := \lim_{\epsilon \searrow 0} \inf \{r(\zeta); 0 < |\zeta - c| < \epsilon\}$), has the property that $-\log \tilde{r}(z_2)$ is a subharmonic function on a neighborhood of 0, because of the subharmonicity of $\frac{1}{j} \log | \sum_{k=0}^{\infty} a_{jk} z_2^k |$ and the Cauchy-Hadamard formula. By the L^2 method, it will turn out that any subharmonic function can be approximated (in an appropriate sense) by a subharmonic function on $D \subset \mathbb{C}$ of the form

$$\log \sum_j |f_j|^2 \quad (f_j \in \mathscr{O}(D))$$

(cf. Chap. 4).

Cauchy's formula holds because holomorphic functions locally admit primitives, but Stokes' formula says that (1.2) holds as well if f is of class C^1 on $\overline{D_1 \times \cdots \times D_n}$ and satisfies the Cauchy-Riemann equation

$$\bar{\partial} f := \sum_{j=1}^{n} \frac{\partial f}{\partial \bar{z}_j} d\bar{z}_j = 0 \quad on \ D_1 \times \cdots \times D_n. \tag{1.3}$$

Here

$$\frac{\partial}{\partial \bar{z}_j} = \frac{1}{2} \left(\frac{\partial}{\partial x_j} + i \frac{\partial}{\partial y_j} \right) \quad and \ d\bar{z}_j = dx_j - i dy_j.$$

Hence, as is well known, any C^1 function satisfying the Cauchy–Riemann equation is holomorphic. The following characterization of holomorphic functions is equally important for later purposes.

Theorem 1.2. *Let f be a measurable function on an open set $U \subset \mathbb{C}^n$ which is square integrable on every compact subset of U with respect to the Lebesgue measure $d\lambda (= d\lambda_n)$. Suppose that*

$$\int_U f \cdot \frac{\partial \phi}{\partial \bar{z}_j} d\lambda = 0 \quad for \ all \ j$$

holds for any $\mathbb{C}$-valued C^∞ function ϕ on U whose support is a compact subset of U. Then f is almost everywhere equal to a holomorphic function on U.

The proof of Theorem 1.2 is done by approximating f locally by taking convolutions with radially symmetric smooth functions with compact support. The same method works to characterize holomorphic functions as those distributions which are weak solutions of the Cauchy–Riemann equation.

1.1.2 Weierstrass Preparation Theorem

In a paper of K. Weierstrass published in 1879, the following is proved.

Theorem 1.3. *Let $F(z_1, \ldots, z_n)$ be a holomorphic function on a neighborhood of the origin $(0, \ldots, 0)$ of $\mathbb{C}^n$ satisfying $F(0, \ldots, 0) = 0$ and $F_0(z_1) := F(z_1, 0, \ldots, 0) \not\equiv 0$. Let p be the integer such that $F_0(z_1) = z_1^p G(z_1), G(0) \neq 0$. Then there exist a holomorphic function of the form*

$$z_1^p + a_1 z_1^{p-1} + \cdots + a_p$$

say $f(z_1; z_2, \ldots, z_n)$, where a_k are holomorphic functions in $(z_2, \ldots, z_n)$ satisfying $a_k(0, \ldots, 0) = 0$, and a function $g(z_1, \ldots, z_n)$ holomorphic and nowhere vanishing in a neighborhood of the origin, such that

$$F = f \cdot g$$

holds in a neighborhood of the origin.

Theorem 1.3 is called the **Weierstrass preparation theorem**. Functions $f(z_1; z_2, \ldots, z_n)$ are called **Weierstrass polynomials** in z_1. Weierstrass polynomials are also called **distinguished polynomials** because they are polynomials in z_1 with "distinguished coefficients" in the ring $\mathbb{C}\{z'\}$. Since the elements of $\mathbb{C}\{z - a\}$ ($a \in \mathbb{C}^n$) are the building blocks of holomorphic functions, the algebraic structures of $\mathbb{C}\{z - a\}$ and their relations to those of $\mathbb{C}\{z - b\}$ for nearby b are particulary important in the local theory of holomorphic functions. The Weierstrass preparation theorem is the most basic tool for studying such properties of the convergent power series rings. There are several proofs of Theorem 1.3 including purely algebraic ones (cf. [Ng, p. 191]), but Cauchy's integral formula gives a very straightforward one:

Proof of Theorem 1.3. By assumption, there exist a neighborhood U_1 of $0 \in \mathbb{C}$ and a neighborhood U of the origin of $\mathbb{C}^{n-1}$ such that, for any $z' \in U$ one can find $s_1, \ldots, s_p \in \mathbb{C}$ satisfying

$$F(z_1, z') = 0 \text{ and } (z_1, z') \in U_1 \times U \iff z_1 \in \{s_1, \ldots, s_p\}.$$

Hence it suffices to show that

$$f(z_1; z') := (z_1 - s_1) \cdots (z_1 - s_p) = z_1^p + a_1(z')z_1^{p-1} + \cdots + a_p(z')$$

is holomorphic. By Cauchy's integral formula,

$$s(m) := s_1^m + \cdots + s_p^m = \frac{1}{2\pi i} \int_{|\zeta|=\epsilon} \frac{\zeta^m}{F(\zeta, z')} \frac{\partial F(\zeta, z')}{\partial \zeta} d\zeta$$

holds for $m \in \mathbb{N}$ and for sufficiently small $\epsilon > 0$. Hence $s(m)$ is holomorphic in z'. By Newton's identity

$$s(m) = a_1 s(m-1) - a_2 s(m-2) + \cdots + (-1)^{m-1} m a_m \quad (1 < m \leq p).$$

Hence a_j is a polynomial in $s_1, \ldots, s_j$, so that f is holomorphic. $\qquad\square$

From now on, the origin $(0, \ldots, 0)$ will be denoted simply by 0. In a paper published in 1887, L. Stickelberger wrote the following as a lemma.

Theorem 1.4. *Let F and G be as in Theorem 2.1. Then, for any $g \in \mathbb{C}\{z\}$ one can find $q \in \mathbb{C}\{z\}$ and $h \in \mathbb{C}\{z'\}[z_1]$ such that $\deg_{z_1} h \leq p - 1$ and $g = qF + h$.*

Proof. By Theorem 1.3, it suffices to show the assertion when F is a Weierstrass polynomial. Let us put

$$q(z_1, z') = \frac{1}{2\pi i} \int_{|\zeta - z_1|=\epsilon} \frac{g(\zeta, z')d\zeta}{F(\zeta, z')(\zeta - z_1)}.$$

Here ϵ is sufficiently small and $\|z'\| \ll \epsilon$. Then q is holomorphic in a neighborhood of 0. Letting

$$h(z_1, z') = \frac{1}{2\pi i} \int_{|\zeta - z_1| = \epsilon} \frac{F(\zeta, z') - F(z_1, z')}{\zeta - z_1} \frac{g(\zeta, z')d\zeta}{F(\zeta, z')},$$

one has $h \in \mathbb{C}\{z'\}[z_1]$, $\deg_{z_1} h \leq p - 1$ and $g = qF + h$.   □

For simplicity, we put $\mathbb{C}\{z - a\} = \mathcal{O}_a$.

Definition 1.1. An invertible element f of $\mathcal{O}_a$ is called a **unit**. f is called a **prime element** if it is not the product of two elements $g, h \in \mathcal{O}_a$ which are both not units. Two elements of $\mathcal{O}_a$ say f and g are said to be **relatively prime** to each other if there exist no $h \in \mathcal{O}_a$ with $h(a) = 0$ dividing both f and g.

The following theorems, which are essentially corollaries of Theorem 1.4, carry the flavor of Euclid's $\Sigma TOIXEIA$.

Theorem 1.5. $\mathcal{O}_a$ *is a unique factorization domain.*

Here a commutative ring with the multiplicative identity and without zero divisors is called a **unique factorization domain** if every nonzero element is decomposed into the product of prime elements uniquely up to multiplication of units.

Theorem 1.6. *Let D be a domain in $\mathbb{C}^n$ and let $f, g \in \mathcal{O}(D)$. If the germs of f and g are relatively prime to each other at a point $c \in D$, then so are they at all points in a neighborhood of c.*

Theorem 1.4 is called the **Weierstrass division theorem**. As one of its important applications, let us mention the following.

Theorem 1.7. *Let D be a domain in $\mathbb{C}^n$ and let $F \in \mathcal{O}(D) \setminus \{0\}$. Then, for every point $c \in F^{-1}(0)$, there exist a neighborhood $U \ni c$ and $f \in \mathcal{O}(U)$ such that, for any $d \in U$ and for any $g \in \mathcal{O}_d$ vanishing along $F^{-1}(0)$ on a neighborhood of d, f_d divides g.*

The set $F^{-1}(0)$ is called a (complex) **hypersurface** of D and f as above is called a **minimal local defining function** of $F^{-1}(0)$.

  In view of these classical theorems, a natural question is to extend them to vector–valued holomorphic functions. Namely, what can we say about the local defining functions of the common zeros of holomorphic functions? An answer was given by the Oka–Cartan theory which will be reviewed in the last section of this chapter. In Chap. 3, refinements of Oka–Cartan theory by the L^2 method will be given following the development in the recent decades. For that the following is important as well as Theorem 1.2. For the proof, see [Oh-21, Proposition 1.14] for instance.

Theorem 1.8. *Let D be a domain in $\mathbb{C}^n$, let $F \in \mathcal{O}(D) \setminus \{0\}$, and let $g \in \mathcal{O}(D \setminus F^{-1}(0))$. If*

$$\int_{D \setminus F^{-1}(0)} |g(z)|^2 d\lambda < \infty,$$

then g is holomorphically extendible to D, i.e., there exists $\tilde{g} \in \mathcal{O}(D)$ such that $\tilde{g}|_{D \setminus F^{-1}(0)} = g$.

In Chap. 3, the reader will find a relation between subharmonicity and the Weierstrass division theorem bound by the L^2 theory (cf. Theorem 3.19).

1.1.3   Domains of Holomorphy and Plurisubharmonic Functions

In view of the starting point that a holomorphic function is a collection of elements of $\mathcal{O}_c$ $(c \in \mathbb{C}^n)$, it is natural to extend the class of domains in $\mathbb{C}^n$ to the domains *over* $\mathbb{C}^n$.

Definition 1.2. A **domain over a topological space** X is a connected topological space $\tilde{X}$ with a local homeomorphism $p : \tilde{X} \to X$.

$\tilde{X}$ is said to be **finitely sheeted** if the cardinality of $p^{-1}(c)$ is bounded from above by some $m \in \mathbb{N}$. A domain D in $\mathbb{C}^n$ is naturally identified with a domain over $\mathbb{C}^n$ with respect to the inclusion map. A domain over X will be referred to also as a **Riemann domain** over X. For two domains $(\mathcal{D}_k, p_k)$ $(k = 1, 2)$ over $\mathbb{C}^n$, $(\mathcal{D}_1, p_1)$ is called a **subdomain** of $(\mathcal{D}_2, p_2)$ if there exists an injective local homeomorphism $\iota : \mathcal{D}_1 \to \mathcal{D}_2$ such that $p_2 \circ \iota = p_1$. Let $\mathcal{D}$ be a domain over $\mathbb{C}^n$. A $\mathbb{C}$–valued function f on $\mathcal{D}$ is said to be **holomorphic** if every point $x \in \mathcal{D}$ has a neighborhood U such that $p|_U$ is a homeomorphism and $f \circ (p|_U)^{-1} \in \mathcal{O}(p(U))$. The set of holomorphic functions on $\mathcal{D}$ will be denoted by $\mathcal{O}(\mathcal{D})$. For any $c \in \mathbb{C}^n$ and for any $f \in \mathcal{O}_c$, a pair of a domain $(\mathcal{D}, p)$ over $\mathbb{C}^n$ and $f_{\mathcal{D}} \in \mathcal{O}(\mathcal{D})$ is called an **extension** of f if there exist $\tilde{c} \in \mathcal{D}$ with $p(\tilde{c}) = c$ and a neighborhood $U \ni \tilde{c}$ such that f is the germ of $f_{\mathcal{D}} \circ (p|_U)^{-1}$ at c. Extensions of f are ordered by the inclusion relation defined as above.

Definition 1.3. A domain $(\mathcal{D}, p)$ over $\mathbb{C}^n$ is called a **domain of holomorphy** if it is the domain of definition of the maximal extension of an element of $\mathcal{O}_{p(c)}$ for some $c \in \mathcal{D}$.

Example 1.1. $(\mathbb{C}^n, id_{\mathbb{C}^n})$ is a domain of holomorphy.

In contrast to this trivial example, the following is highly nontrivial.

Theorem 1.9. *Every domain over $\mathbb{C}$ is a domain of holomorphy.*

For the proof, the idea of Oka for the characterization of domains of holomorphy for any n is essentially needed. (See [B-S] or Chap. 2.)

Theorem 1.1 shows that not every domain in $\mathbb{C}^n$ is a domain of holomorphy if $n \geq 2$. Accordingly, the classification theory of holomorphic functions can be geometric, with respect to the Euclidean distance

$$\text{dist}(z, w) := \|z - w\| = \left(\sum_{j=1}^{n} |z_j - w_j|^2 \right)^{\frac{1}{2}}, \quad z, w \in \mathbb{C}^n$$

for instance. The convexity notion is important in this context as one can see from:

Theorem 1.10. *For a domain $\Omega \subset \mathbb{R}^n$, the domain $\{z \in \mathbb{C}^n; \text{Re} z \in \Omega\}$ is a domain of holomorphy if and only if Ω is convex.*

Sketch of Proof. If Ω is convex and $x_0 \in \partial\Omega$, there exists an affine linear function ℓ on $\mathbb{C}^n$ such that $\ell(\mathbb{R}^n) = \mathbb{R}$, $\ell(x_0) = 0$ and $\ell(x) > 0$ if $x \in \Omega$. Then

$$\frac{1}{\ell(z + iy_0)} \in \mathscr{O}(\{\text{Re} z \in \Omega\})$$

for any $y_0 \in \mathbb{R}^n$. Hence, considering an infinite sum of such functions, it is easy to see that $\{\text{Re} z \in \Omega\}$ is a domain of holmorphy. For the converse, see [Hö-2, Theorem 2.5.10].   □

The convexity notion is naturally attached to $\sigma \in \mathbb{C}\{z\}$ as follows. Let

$$R_\sigma := \{(z_1, \ldots, z_n); \sigma \text{ is convergent at } (z_1, \ldots, z_n)\}^\circ.$$

For any domain $\Omega \subset \mathbb{C}^n$ satisfying

$$\Omega = \{(\zeta_1 z_1, \ldots, \zeta_n z_n); (z_1, \ldots, z_n) \in \Omega \text{ and } \zeta_j \in \mathbb{D}\},$$

we put $\log|\Omega| = \{(\log r_1, \ldots, \log r_n) \in [-\infty, \infty)^n; (r_1, \ldots, r_n) \in \Omega\}$. Then it is easy to see that the set $\log|R_\sigma|$ is convex for any σ. Conversely, $\Omega = R_\sigma$ for some $\sigma \in \mathbb{C}\{z\}$ if $\log|\Omega|$ is convex (cf. [Oh-21, Corollary 1.19]). R_σ is called the **Reinhardt domain** of σ.

On the other hand, the observation after Theorem 1.1 can be stated as follows.

Theorem 1.11. *Let $D \subset \mathbb{C}$ be a domain, let φ be an upper semicontinuous function on D, and let $\tilde{D} = \{z = (z_1, z_2) \in \mathbb{C}^2; |z_1| < e^{-\varphi(z_2)}\}$. If $\tilde{D}$ is a domain of holomorphy, then $\varphi(z_2)$ is subharmonic on D.*

Definition 1.4. For any Riemann domain $p : \mathscr{D} \to \mathbb{C}^n$, a function $\varphi : \mathscr{D} \to [-\infty, \infty)$ is called a **pseudoconvex function** on $\mathscr{D}$ if every point $x_0 \in \mathscr{D}$ admits a neighborhood U such that $\{(x, \zeta) \in U \times \mathbb{C}; |\zeta| < e^{-\varphi(x)}\}$ is a domain of holomorphy over $\mathbb{C}^n$.

By Theorem 1.11, for any pseudoconvex function $\varphi : \mathscr{D} \to [-\infty, \infty)$ and for any complex line $\ell \subset \mathbb{C}^n$, $\varphi|_{p^{-1}(\ell)}$ is subharmonic with respect to complex coordinates on ℓ.

Definition 1.5. For any Riemann domain $p : \mathscr{D} \to \mathbb{C}^n$, a function $\varphi : \mathscr{D} \to [-\infty, \infty)$ is called a **plurisubharmonic function** on $\mathscr{D}$ if $\varphi|_{p^{-1}(\ell)}$ is subharmonic with respect to complex coordinates on ℓ for any complex line $\ell \subset \mathbb{C}^n$.

Given $p : \mathscr{D} \to \mathbb{C}^n$, the most basic property of the domains of holomorphy is described in terms of the function

$$\delta_{\mathscr{D}}(x) = \sup \{r; p \text{ maps a neighborhood of } x \text{ bijectively to } \mathbb{B}^n(p(x), r)\},$$

where $\mathbb{B}^n(c, r) := \{z \in \mathbb{C}^n; \|z - c\| < r\}$. This "distance from x to $\partial\mathscr{D}$" satisfies the following remarkable property.

Theorem 1.12 (Oka's lemma). *Let $\mathscr{D}$ be a domain of holomorphy over $\mathbb{C}^n$. Then $-\log \delta_{\mathscr{D}}$ is a plurisubharmonic function.*

For the proof of Oka's lemma, see [H, G-R] or [Oh-2].

A very profound fact in several complex variables is that the converse of Oka's lemma is true. As a result, it follows that every plurisubharmonic function is pseudoconvex. An approach to this by the L^2 method is one of the main objects of the discussions in the subsequent chapters.

1.2 Complex Manifolds and Convexity Notions

Plurisubharmonic functions play an important role in the study of basic existence problems. This is also the case on complex manifolds, objects on which the theory of holomorphic mappings can be discussed in full generality. In the study of global coordinates on complex manifolds, analytic tools available on the domains over $\mathbb{C}^n$ work as well on those manifolds that satisfy certain convexity properties. Basic convexity notions needed in this theory are recalled and important existence theorems due to Oka and Grauert will be reviewed.

1.2.1 Complex Manifolds, Stein Manifolds and Holomorphic Convexity

By a **complex manifold**, we shall mean a Hausdorff space M with a countable basis of open sets and with an open covering $\{U_j\}_{j \in \mathbb{I}}$, for some index set $\mathbb{I}$, such that a homeomorphism φ_j from a domain D_j in $\mathbb{C}^n$ ($n = n(j)$) to U_j is attached for each j, in such a way that $\varphi_j^{-1} \circ \varphi_k$ is holomorphic on $\varphi_k^{-1}(U_j \cap U_k)$ whenever

$U_j \cap U_k \neq \emptyset$. Unless otherwise stated, every connected component of M is assumed to be paracompact, i.e. admits a countable basis of open sets. (In most cases M is tacitly assumed to be connected.) (U_j, φ_j^{-1}) is called a **chart** of M and the collection $\{(U_j, \varphi_j^{-1})\}_{j \in \mathbb{I}}$ of charts is called an **atlas** of M. A $\mathbb{C}$-valued function f on M is said to be holomorphic if $f \circ \varphi_j$ are holomorphic on D_j. The set of holomorphic functions on M is denoted by $\mathcal{O}(M)$. The set of germs of holomorphic functions at x will be denoted by $\mathcal{O}_{M,x}$. For any local coordinate z around x, one has an isomorphism $\mathcal{O}_{M,x} \cong \mathbb{C}\{z\}$. Unless stated otherwise, atlases are taken to be maximal with respect to the inclusion relation. φ_j^{-1} are called **local coordinates** around $x \in U_j$ and $\varphi_j^{-1} \circ \varphi_k$ are called **coordinate transformations**. By an abuse of language, for a local coordinate ψ around a fixed point $x \in M$, the condition $\psi(x) = 0$ will be assumed tacitly in many cases. A complex manifold M is said to be **of dimension** n if $\max_j \dim D_j = n$ and **of pure dimension** n if $\dim D_j = n$ for all j. Unless stated otherwise, complex manifolds will be assumed to be finite dimensional and of pure dimension. By an abuse of language, topological spaces with discrete topology are regarded as 0-dimensional complex manifolds. It is conventional to call connected 1-dimensional complex manifolds **Riemann surfaces**. $\mathbb{C}^n$ and the domains over $\mathbb{C}^n$ are regarded as a complex manifold in an obvious way.

Example 1.2. Let $D_j = \mathbb{C}^n$ $(j = 0, 1, \ldots, n)$ and let

$$\mathbb{CP}^n = \left(\coprod_{j=0}^{n} D_j \right) / \sim,$$

where $\sim$ is an equivalence relation defined by

$$D_j \ni (z_1, \ldots, z_n) \sim (w_1, \ldots, w_n) \in D_k$$

$$\Longleftrightarrow$$

$$(z_1, \ldots, z_j, 1, z_{j+1}, \ldots, z_n) // (w_1, \ldots, w_k, 1, w_{k+1}, \ldots, w_n).$$

Here $v // w$ means that there exists $\zeta \in \mathbb{C} \setminus \{0\}$ such that $\zeta v = w$.

Then, with respect to the quotient topology and the natural maps $\varphi_j : D_j \to \mathbb{CP}^n$ induced from the inclusion, $\mathbb{CP}^n$ is (or rather *becomes*, more precisely speaking) a compact complex manifold. $\mathbb{CP}^n$ is called the **complex projective space** of dimension n.

Let M_μ $(\mu = 1, 2)$ be two complex manifolds with atlases $\{(U_{\mu,j}, \phi_{\mu,j}^{-1})\}_{j \in \mathbb{I}_\mu}$, respectively. Then the product space $M_1 \times M_2$ is a complex manifold with respect to a (non-maximal) atlas $\{(U_{1,j} \times U_{2,k}, (\phi_{1,j}, \phi_{2,k})^{-1})\}_{(j,k) \in \mathbb{I}_1 \times \mathbb{I}_2}$. A continuous map F from M_1 to M_2 is called a **holomorphic map** if $\varphi_{2,k}^{-1} \circ F \circ \varphi_{1,j}$ are all holomorphic. The set of holomorphic maps from M_1 to M_2 will be denoted by $\mathcal{O}(M_1, M_2)$.

Given a surjective holomorphic map $f : M_1 \to M_2$, a map s from an open set U in M_2 to M_1 will be called a **section** if $f \circ s = id_U$ holds. (s need not be holomorphic.) A holomorphic map F is said to be **biholomorphic** if it has a holomorphic inverse. If $\mathcal{O}(M_1, M_2)$ contains a biholomorphic map, M_1 and M_2 are said to be **isomorphic** to each other (denoted by $M_1 \cong M_2$). Aut$M$ will stand for the group of biholomorphic automorphisms of M. A proper holomorphic map $F : M_1 \to M_2$ is called **almost biholomorphic** if there exists a nowhere–dense subset A of M_1 such that $F|_{M_1 \setminus A}$ is a biholomorphic map onto its image.

Example 1.3. Let $\pi : \mathbb{C}^{n+1} \setminus \{0\} \to \mathbb{CP}^n$ be defined by

$$\pi(\xi_0, \xi_1, \ldots, \xi_n) = \phi_j(z_1, \ldots, z_n) \quad \text{for} \quad \xi_j \neq 0$$

and

$$(\xi_0, \xi_1, \ldots, \xi_n) // (z_1, \ldots, z_{j-1}, 1, z_j, \ldots, z_n).$$

Then $\pi \in \mathcal{O}(\mathbb{C}^{n+1} \setminus \{0\}, \mathbb{CP}^n)$. $\xi = (\xi_0, \xi_1, \ldots, \xi_n)$ is called the **homogeneous coordinate** of $\mathbb{CP}^n$. $\pi(\xi_0, \xi_1, \ldots, \xi_n)$ will be denoted by $[(\xi_0, \ldots, \xi_n)]$ (the equivalence class) or $(\xi_0 : \xi_1 : \cdots : \xi_n)$ (continued ratio). For any ($\mathbb{C}$-vector) subspace $V \subset \mathbb{C}^{n+1}$ of codimension one, $\pi(V \setminus \{0\})$ is called a **complex hyperplane**.

A holomorphic map $F : M_1 \to M_2$ is called an **embedding** if the following are satisfied:

(1) F is injective.
(2) For any point $p \in M_1$ there exist a neighborhood $U \ni p$ and a chart (V, ψ) of M_2 such that $F(p) \in V$, $\psi(F(p)) = 0$ and

$$F(U) = \{q \in V; \psi(q)_1 = \cdots = \psi(q)_k = 0\} \quad \text{for some} \ k = k(p).$$

The image $F(M_1)$, equipped with the topology of M_1, of a holomorphic embedding F will be called a **complex submanifold** of M_2. The integer $\min_{p \in M_1} k(p)$ is called the **codimension** of $F(M_1)$. By an abuse of language, we shall call $F(M_1)$ a closed complex submanifold of M if F is a proper holomorphic embedding. A **closed complex submanifold** of an open subset of M is called a **locally closed** complex submanifold of M. Closed submanifolds of codimension one are called **complex hypersurfaces**.

A holomorphic embedding from $\mathbb{D}$ to M is called a **holomorphic disc** in M. An upper semicontinuous function $\Phi : M \to [-\infty, \infty)$ is called a **plurisubharmonic function** on M if $\Phi \circ \iota$ is subharmonic for any holomorphic disc $\iota : \mathbb{D} \hookrightarrow M$. It is easy to see that Φ is plurisubharmonic if and only if $\Phi \circ \psi^{-1}$ is plurisubharmonic for every chart (U, ψ) in the sense of Definition 1.5.

Given any subgroup $G \subset \text{Aut}M$ such that

(1) $\gamma \cdot x := \gamma(x) \neq x$ if $G \ni \gamma \neq id_M$
(2) $\#\{\gamma \in G; \ \gamma(K) \cap K \neq \varnothing\} < \infty$ for any compact set $K \subset M$,

where $\sharp A :=$ the cardinality of A, the projection $\pi : M \to M/G := \{G \cdot x \, ; \, x \in M\}$ naturally induces on M/G a complex manifold structure.

Example 1.4 (complex semitori). Let Γ be an additive subgroup of $\mathbb{C}^n$ of the form $\sum_{j=1}^{m} \mathbb{Z} \cdot v_j$ $(v_j \in \mathbb{C}^n)$ such that $v_1, v_2, \ldots, v_m$ are linearly independent over $\mathbb{R}$. Then Γ is naturally identified with a subgroup of $\mathrm{Aut}(\mathbb{C}^n)$ by

$$\Gamma \ni v \mapsto \{z \mapsto z + v\} \in \mathrm{Aut}\mathbb{C}^n.$$

Since (1) and (2) are obviously satisfied by Γ, one has a complex manifold $\mathbb{C}^n/\Gamma$ which is called a **complex semitorus**. $\mathbb{C}^n/\Gamma$ is called a **complex torus** if it is compact, or equivalently $m = 2n$. A well–known theorem of Riemann says that a complex torus $\mathbb{C}^n/\Gamma$ can be embedded holomorphically into $\mathbb{CP}^{2n+1}$ if

$$(v_1, v_2, \ldots, v_{2n}) = (I, Z)$$

holds for the $n \times n$ identity matrix I and an $n \times n$ symmetric matrix Z whose imaginary part is positive definite. Here v_j are identified with the corresponding column vectors.

Clearly, complex semitori are pseudoconvex. From this viewpoint, a generalization of Riemann's theorem by Kodaira will be discussed in Chap. 2 as an application of the L^2 method.

Example 1.5. The map

$$\iota : \mathbb{C}^{n+1} \setminus \{0\} \ni \xi \mapsto (\xi, [\xi]) \in \mathbb{C}^{n+1} \times \mathbb{CP}^n$$

is a holomorphic embedding. The closure of $\iota(\mathbb{C}^{n+1} \setminus \{0\})$ is a closed complex submanifold. The restriction of the projection $\mathbb{C}^{n+1} \times \mathbb{CP}^n \to \mathbb{C}^{n+1}$ to $\overline{\iota(\mathbb{C}^{n+1} \setminus \{0\})}$, say ϖ, is a surjective holomorphic map which is almost biholomorphic. Note that $\varpi^{-1}(0) \cong \mathbb{CP}^n$. ϖ is called the blow–up centered at $0 \in \mathbb{C}^{n+1}$. Blow–ups centered at (or along) closed complex submanifolds are defined similarly.

Compact complex manifolds which are isomorphic to closed complex submanifolds of $\mathbb{CP}^n$ are called **projective algebraic manifolds** (over $\mathbb{C}$). A theorem of Chow [Ch] says that every projective manifold is the set of zeros of some homogeneous polynomial in ξ. It may be worthwhile to mention that Chow's theorem is a corollary of a continuation theorem of Hartogs type (cf. [R-S]).

Example 1.6 (Hopf manifolds). Let $\mathscr{H} = (\mathbb{C}^n \setminus \{0\})/ \sim$, where $\sim$ is an equivalence relation defined by

$$(z_1, \ldots, z_n) \sim (w_1, \ldots, w_n) \iff w_k = e^m \cdot z_k \ (1 \le k \le n) \text{ for some } m \in \mathbb{Z}.$$

Then, with respect to the quotient topology and the restrictions of the canonical projection $p : \mathbb{C}^n \setminus \{0\} \to \mathscr{H}$ to the domains D such that $p|_D$ is injective, $\mathscr{H}$ becomes a compact complex manifold. By applying a continuation theorem of Hartogs type, or appealing to the fact that the p-th Betti numbers of projective algebraic manifolds are even integers if p is odd, one knows that $\mathscr{H}$ is not projective algebraic.

Definition 1.6. A **Stein manifold** is a complex manifold M such that any closed discrete subset of M is mapped bijectively to some closed discrete subset of $\mathbb{C}$ by some element of $\mathcal{O}(M)$.

Theorem 1.13 (cf. [Bi, R-1, N]). *A complex manifold M of dimension n is Stein if and only if there exists a proper holomorphic embedding from M to $\mathbb{C}^{2n+1}$.*

Remark 1.1. It is known that Stein manifolds of dimension n are properly and holomorphically embeddable into $\mathbb{C}^{[\frac{3n}{2}]+1}$ if $n \geq 2$ (cf. [E-Grm, Sm]).

Definition 1.7. A complex manifold M is said to be **holomorphically convex** if any closed discrete subset of M is properly mapped onto some closed discrete subset of $\mathbb{C}$ by some element of $\mathcal{O}(M)$.

Theorem 1.14 (cf. [Gra-1]). *An n–dimensional complex manifold M is Stein if and only if the following are satisfied:*

(1) M is holomorphically convex.
(2) For any two distinct points $p, q \in M$, there exists $f \in \mathcal{O}(M)$ such that $f(p) \neq f(q)$.
(3) For any $p \in M$ there exist a neighborhood $U \ni p$ and $f_1, \ldots, f_n \in \mathcal{O}(M)$ such that $(U, (f_1, \ldots, f_n))$ is a chart of M.

Remark 1.2. The class of Stein manifolds was first introduced by K. Stein in [St] by the properties *(1)* $\sim$ *(3)* as above. So, Definition 1.6 was a characterization of Stein manifolds.

Grauert also established another characterization of Stein manifolds by generalizing Oka's theory on pseudoconvex domains over $\mathbb{C}^n$, which will be reviewed in Sect. 1.2.3 after a preliminary in Sect. 1.2.2

1.2.2 Complex Exterior Derivatives and Levi Form

Let us recall that differentiable manifolds of class C^r, for $0 \leq r \leq \infty$ or $r = \omega$, are defined by replacing the domains D_j in $\mathbb{C}^n$ by domains in $\mathbb{R}^m$ and requiring $\varphi_j^{-1} \circ \varphi_k$ to be of class C^r. Basic terminology on differentiable manifolds such as C^r maps, tangent bundles, differential forms, exterior derivatives, etc. will be used freely (cf. [W]). By an abuse of notation, $C^r(M)$ will stand for the set of $\mathbb{C}$-valued C^r functions on M. The set of germs of $\mathbb{C}$-valued C^r functions at x will be denoted by $\mathscr{C}_{M,x}^r$.

Let M be a complex manifold of dimension n. By $T_M^{\mathbb{C}}$ we shall denote the complex tangent bundle of M, i.e. the complexification of the tangent bundle T_M of M as a differentiable manifold. Recall that

$$T_M^{\mathbb{C}} = \coprod_{x \in M} T_{M,x}^{\mathbb{C}}$$

as a set, where

$$T_{M,x}^{\mathbb{C}} = \{v \in \mathrm{Hom}(\mathscr{C}_{M,x}^{\infty}, \mathbb{C}); v(fg) = f(x)v(g) + g(x)v(f)\}.$$

Here $\mathrm{Hom}(A, B)$ denotes the set of $\mathbb{C}$ linear maps from A to B.

Let

$$(T_M^{\mathbb{C}})^* = \coprod_{x \in M} (T_{M,x}^{\mathbb{C}})^* \quad (V^* = \mathrm{Hom}(V, \mathbb{C}) \textit{ for any complex vector space } V)$$

be the complex cotangent bundle of M, i.e. the dual bundle of $T_M^{\mathbb{C}}$. For any $x \in M$ we put

$$T_{M,x}^{0,1} = \{v \in T_{M,x}^{\mathbb{C}}; v(f) = 0 \text{ if } f \in \mathscr{O}_{M,x}\},$$

$$T_{M,x}^{1,0} = \overline{T_{M,x}^{0,1}} \quad (\textit{complex conjugate})$$

and

$$T_M^{0,1} = \coprod_{x \in M} T_{M,x}^{0,1}, \qquad T_M^{1,0} = \coprod_{x \in M} T_{M,x}^{1,0}.$$

$T_M^{1,0}$ is called the **holomorphic tangent bundle** of M. We put

$$(T_M^{1,0})^* = \coprod_{x \in M} (T_{M,x}^{1,0})^*$$

and

$$(T_M^{p,q})^* = \wedge^p (T_M^{1,0})^* \otimes \wedge^q (T_M^{0,1})^*.$$

Then

$$\wedge^r (T_M^{\mathbb{C}})^* \cong \oplus_{p+q=r} (T_M^{p,q})^*.$$

According to this decomposition, the exterior derivative d acting on the set of C^{∞} sections of $\wedge^r (T_M^{\mathbb{C}})^*$ decomposes naturally into the sum of the complex exterior derivative of type $(1,0)$, denoted by ∂, and its conjugate $\bar{\partial}$, the complex exterior derivative of type $(0,1)$.

In terms of a local coordinate z,

$$du = d(\sum u_{I\bar{J}}dz_I \wedge d\bar{z}_J) = \partial u + \bar{\partial}u,$$

where

$$\partial(\sum_{I,J} u_{I\bar{J}}dz_I \wedge d\bar{z}_J) = \sum_{j,I,J} \frac{\partial u_{I\bar{J}}}{\partial z_j}dz_j \wedge dz_I \wedge d\bar{z}_J$$

and

$$\bar{\partial}(\sum_{I,J} u_{I\bar{J}}dz_I \wedge d\bar{z}_J) = \sum_{j,I,J} \frac{\partial u_{I\bar{J}}}{\partial \bar{z}_j}d\bar{z}_j \wedge dz_I \wedge d\bar{z}_J.$$

For any chart (U, ψ) of M, say $\psi = (z_1, \ldots, z_n)$, there is a natural identification

$$\mathbb{C}^n \longrightarrow T_{M,x}^{0,1} \quad x \in U$$

$$\xi = (\xi^1, \ldots, \xi^n) \mapsto v_x(\xi)$$

by

$$v_x(\xi)(f \circ \psi) = \sum \xi^j \frac{\partial f}{\partial \bar{z}_j}(\psi(x)), \quad f \in C^\infty(\psi(U)).$$

The section $x \mapsto v_x(\xi)$ (resp. $x \mapsto \overline{v_x(\bar{\xi})}$) of $T_M^{0,1}$ (resp. $T_M^{1,0}$) over U will be denoted by $\sum \xi^j \frac{\partial}{\partial \bar{z}_j}$ (resp. $\sum \xi^j \frac{\partial}{\partial z_j}$).

Given a real–valued C^2 function φ on M, the **Levi form** of φ at $x \in M$ is defined as a Hermitian form

$$\sum_{j,k=1}^n \frac{\partial^2 \varphi}{\partial z_j \partial \bar{z}_k} \xi^j \bar{\xi}^k$$

on

$$T_{M,x}^{1,0} \cong \left\{ \sum \xi^j \left(\frac{\partial}{\partial z_j}\right)_{z=\psi(x)} ; \xi \in \mathbb{C}^n \right\}.$$

Although the definition uses $\psi = z$, it is easy to see that the above Hermitian form on $T_{M,x}^{1,0}$ is independent of the choice of local coordinates. The Levi form of φ is denoted simply by $\mathscr{L}_\varphi$, or more explicitly by $\partial\bar{\partial}\varphi$, but by an abuse of notation.

φ is said to be q-**convex** (resp. **weakly** q-**convex**) at x if $\mathscr{L}_\varphi$ has at most $q-1$ nonpositive (resp. negative) eigenvalues at x. It is easy to verify and a fact of basic importance that φ is plurisubharmonic on M if and only if φ is everywhere weakly 1-convex. If φ is 1-convex at x, we shall also say that φ is **strictly plurisubharmonic** at x.

1.2.3   Pseudoconvex Manifolds and Oka–Grauert Theory

A complex manifold M is said to be C^r-pseudoconvex if M admits a C^r plurisubharmonic exhaustion function. Here a real–valued function, say Ψ on a topological space X, is called an **exhaustion function** on X if its sublevel sets

$$X_c = \{x \in X; \Psi(x) < c\}$$

are all relatively compact for all $c < \sup \Psi$. Usually we assume that $\sup \Psi = \infty$ unless Ψ is referred to as a bounded exhaustion function. If M admits a strictly plurisubharmonic exhaustion function, M is called a **1-complete** manifold. M is called q-**convex** if it admits an exhaustion function which is q-convex on the complement of a compact subset of M. It is easy to see that every 1-convex manifold is C^∞-pseudoconvex. In fact, if M admits a C^2 exhaustion function Ψ such that $\mathcal{L}_\Psi$ is positive definite on $M \setminus M_c$, M also admits a C^∞ exhaustion function, say $\tilde{\Psi}$, which is strictly plurisubharmonic outside a compact subset of M. Such a function $\tilde{\Psi}$ is obtained by approximating Ψ by a C^∞ function in the C^2 topology. Then, $\lambda(\tilde{\Psi})$ is a C^∞ plurisubharmonic exhaustion function on M for some C^∞ convex increasing function λ on $\mathbb{R}$. Here, λ is said to be **convex increasing** if $\lambda' \geq 0$ and $\lambda'' \geq 0$. For simplicity, as C^r-pseudoconvex manifolds we shall only consider C^∞-pseudoconvex manifolds. Accordingly, they will be called **pseudoconvex manifolds** from now on. In virtue of Oka's lemma, it is easy to see that locally pseudoconvex domains over $\mathbb{C}^n$ are 1-complete. By an abuse of language, 1-convex manifolds will also be called **strongly pseudoconvex manifolds**. (Apparently this term is better suited to 1-complete manifolds.)

Remark 1.3. A complex manifold M is called a **complex Lie group** if M is equipped with a group structure such that the multiplication is a holomorphic map from $M \times M$ to M. It is known that every complex Lie group is pseudoconvex (cf. [Kz-2]). The notion of q-convexity was first introduced by Rothstein [Rt] in the study of analytic continuation. It also naturally arises in the study of complex homogeneous manifolds (cf. [Huckl]).

Theorem 1.15 (cf. [Gra-3]). *1-complete manifolds are Stein and strongly pseudoconvex manifolds are holomorphically convex.*

For the proof of Theorem 1.15, see [G-R]. A proof by the L^2 method will be given in Chap. 2. Combining Theorems 1.13 and 1.15 one has the following.

Theorem 1.16. *Every real analytic manifold of dimension m is embeddable into $\mathbb{R}^{4m+2}$ by a real analytic map as a closed submanifold.*

Sketch of Proof. Any real analytic manifold, say T, is a closed submanifold of its complexification $T^{\mathbb{C}}$, defined by replacing the local coordinates $(x_1, \ldots, x_m)$ by complex local coordinates $(x_1 + iy_1, \ldots, x_m + iy_m)$. Since $\sum_{k=1}^m y_k^2$ are strictly plurisubharmonic, it is easy to see that T admits a 1-complete neighborhood system in $T^{\mathbb{C}}$. Hence, realizing a neighborhood of T as a closed complex submanifold of $\mathbb{C}^{2m+1}$, we are done.   □

Let $\pi : \Omega \to M$ be a domain over M. Ω is called a **locally pseudoconvex domain over** M if one can find for any $x \in M$ a neighborhood $U \ni x$ such that $\pi^{-1}(U)$ is pseudoconvex.

Theorem 1.17 (Oka–Grauert theorem). *Every locally pseudoconvex domain over a Stein manifold is Stein.*

Corollary 1.1. *Domains of holomorphy over $\mathbb{C}^n$ are Stein.*

Remark 1.4. Corollary 1.1 was first shown by H. Cartan and P. Thullen [C-T], for the domains *in* $\mathbb{C}^n$. Their proof works as well for finitely sheeted domains over $\mathbb{C}^n$. It is remarkable that the generalization to the infinitely sheeted case was established only after Oka's work [O-4] which identified holomorphic convexity with 1-completeness for domains over $\mathbb{C}^n$. The method of Hörmander [Hö-1, Hö-2] affords a quantitative approach to Theorem 1.17.

It is known that every locally pseudoconvex domain over $\mathbb{CP}^n$ is pseudoconvex (cf. Theorem 2.73). As a result, a locally pseudoconvex domain over $\mathbb{CP}^n$ is Stein unless it is biholomorphic to $\mathbb{CP}^n$ itself. In general, when D is a domain with smooth boundary in a complex manifold M, local pseudoconvexity is a property of the Levi form of a function defining the boundary ∂D. To describe the boundary behavior of holomorphic functions, the Levi form of a defining function of ∂D is important. It is basic that local pseudoconvexity of D is characterized by an extrinsic but essentially intrinsic geometric property of ∂D.

Let D be a domain in M. For any $r \geq 1$, D is said to be C^r-**smooth** if there exists a real–valued C^r function say ρ on a neighborhood U of ∂D such that

$$D \cap U = \{z \in U; \rho(z) < 0\}$$

and $d\rho$ vanishes nowhere on ∂D. We shall call ρ a **defining function** of ∂D, or sometimes that of D if ρ is defined on $D \cup U$. We put

$$T_{\partial D}^{1,0} = T_M^{1,0} \cap (T_{\partial D} \otimes \mathbb{C}).$$

If D is C^2-smooth and $\mathscr{L}_\rho|_{T_{\partial D}^{1,0}}$ is everywhere semipositive on ∂D for some defining function ρ of D, ∂D is said to be **pseudoconvex**. ∂D is called **strongly pseudoconvex** at $x \in \partial D$ if $\mathscr{L}_\rho|_{T_{\partial D}^{1,0}}$ is positive definite at x.

Definition 1.8. A **strongly pseudoconvex domain** in M is a relatively compact domain in M whose boundary is everywhere strongly pseudoconvex.

Strongly pseudoconvex domains admit strictly plurisubharmonic defining functions. In fact, for any defining function ρ of D, $e^{A\rho} - 1$ becomes strictly plurisubharmonic on a neighborhood of ∂D for sufficiently large A. Strongly pseudoconvex domains are 1-convex because $-\log(-\rho)$ is an exhaustion function on D which is plurisubharmonic outside a compact subset of D.

Remark 1.5. A smoothly bounded pseudoconvex domain is called **weakly pseudoconvex** if it is not strongly pseudoconvex. There exist C^ω-smooth weakly pseudoconvex domains which do not admit plurisubharmonic defining functions (cf. [B]).

1.3   Oka–Cartan Theory

In order to discuss the questions on the rings and modules of holomorphic functions, it is often necessary to approximate locally defined functions by globally defined ones. The notion of sheaf cohomology serves as a convenient machinery for that purpose.

Once these notions are transplanted from the field of algebraic functions to that of general analytic functions, various new questions naturally arise, because analytic functions show up (to us) not as global objects, but only as local ones. (Kiyoshi Oka—in a letter to Teiji Takagi)

1.3.1   Sheaves and Cohomology

Let $\{\mathscr{F}_x\}_{x\in X}$ be a family of Abelian groups with the identity elements $0_x \in \mathscr{F}_x$ parametrized by a topological space X. Let

$$\mathscr{F} = \coprod_{x\in X} \mathscr{F}_x$$

and let $p : \mathscr{F} \to X$ be defined by $p(\mathscr{F}_x) = \{x\}$. For any open set $U \subset X$, let

$$\mathscr{F}[U] = \{s : U \to \mathscr{F}; p \circ s = id_U\}.$$

By an abuse of language, elements of $\mathscr{F}[U]$ will be called **possibly discontinuous sections** of $\mathscr{F}$. If $s \in \mathscr{F}[U]$ and $s(x) = 0(= 0_x)$ for all $x \in U$, s will be called the **zero section** of $\mathscr{F}$ over U and denoted simply by 0.

Definition 1.9. A family $\{\mathscr{F}(U)\}_U$ of subsets $\mathscr{F}(U)$ of $\mathscr{F}[U]$ is called a **presheaf** if the following are satisfied:

(1) $s \in \mathscr{F}(U), U \supset V \Rightarrow s|_V \in \mathscr{F}(V)$.
(2) $f \in \mathscr{F}_x \Rightarrow$ there exists a neighborhood $U \ni x$ and $s \in \mathscr{F}(U)$ satisfying $s(x) = f$.

(3) $s \in \mathscr{F}(U), x \in U, s(x) = 0_x \Rightarrow s = 0$ on a neighborhood of x.
(4) $s \in \mathscr{F}(U), t \in \mathscr{F}(V) \Rightarrow (s - t)|_{U \cap V} \in \mathscr{F}(U \cap V)$.

A presheaf $\{\mathscr{F}(U)\}_U$ induces a topology on the set $\mathscr{F}$ in such a way that $\cup_{U \subset X}\{s(U); s \in \mathscr{F}(U)\}$ is a basis of open sets of $\mathscr{F}$. Elements of $\mathscr{F}$ are continuous with respect to this topology.

Definition 1.10. A presheaf $\{\mathscr{F}(U)\}$ is called a **sheaf** if $\mathscr{F}(U) = \{s \in \mathscr{F}[U];$ For any $x \in U$ there exists a neighborhood $V \ni x$ such that $s|_V \in \mathscr{F}(V)\}$.

Clearly, for any presheaf $\{\mathscr{F}(U)\}$ (an abbreviation for $\{\mathscr{F}(U)\}_U$), one can find a sheaf $\{\overline{\mathscr{F}(U)}\}$ such that $\mathscr{F}(U) \subset \overline{\mathscr{F}(U)} \subset \mathscr{F}[U]$ uniquely. $\{\overline{\mathscr{F}(U)}\}$ will be called the **sheafification** of $\{\mathscr{F}(U)\}$.

For simplicity, the topological space $\mathscr{F}$ will also stand for the sheaf $\{\mathscr{F}(U)\}$. To be explicit, $\mathscr{F}$ is called a sheaf over X. The map $p : \mathscr{F} \to X$ will be referred to as a **sheaf projection**.

$\mathscr{F}_x$ is called a **stalk** of $\mathscr{F}$ at x, and the elements of $\mathscr{F}_x$ the **germs** at x. Elements of $\mathscr{F}(U)$ will be called the **sections** of $\mathscr{F}$ over U. By (3) above, the germs at x of sections in $\mathscr{F}(U)$ are naturally identified with elements of $\mathscr{F}_x$ if $U \ni x$, i.e.

$$\mathscr{F}_x = \mathrm{ind.lim}_{U \ni x} \mathscr{F}(U)$$

with respect to the inductive system induced from the natural restriction maps $\mathscr{F}[U] \to \mathscr{F}[V]$ for $U \supset V \ni x$. For any $s \in \mathscr{F}[U]$ the germ of s at x will be denoted by s_x. In short, $s(x) = s_x$ if $s \in \mathscr{F}(U)$. Let $\mathscr{G}$ be another sheaf over X. $\mathscr{G}$ is called a **subsheaf** of $\mathscr{F}$ if $\mathscr{G}(U) \subset \mathscr{F}(U)$ for any open set $U \subset X$.

Note that the family $\{\mathscr{F}[U]\}$ itself is not necessarily a sheaf because the condition (3) may not be satisfied. However, if we put

$$\hat{\mathscr{F}}_x = \mathrm{ind.lim}_{U \ni x} \mathscr{F}[U],$$

$$\hat{\mathscr{F}} = \coprod_{x \ni X} \hat{\mathscr{F}}_x$$

and

$$\hat{\mathscr{F}}(U) = \{\hat{s} : U \to \hat{\mathscr{F}}; \hat{s}(x) = s_x \text{ for some } s \in \mathscr{F}[U]\},$$

then $\{\hat{\mathscr{F}}(U)\}_U$ is a sheaf over X. $\hat{\mathscr{F}}$ has a property that any section over any open set extends to X as a section. Sheaves having this property are called **flabby sheaves**. Since $\mathscr{F}$ is a subsheaf of $\hat{\mathscr{F}}$, we shall call the sheaf $\hat{\mathscr{F}}$ the **canonical flabby extension** of $\mathscr{F}$.

For any two sheaves $\mathscr{F}_j$ $(j = 1, 2)$ over X, the **direct sum** $\mathscr{F}_1 \oplus \mathscr{F}_2$ is a sheaf defined by $\{\mathscr{F}_1(U) \oplus \mathscr{F}_2(U)\}_U$. For any continuous map $\beta : X \to Y$, the **direct image sheaf** of $\mathscr{F}$ by β, denoted by $\beta_* \mathscr{F}$, is defined over Y by

$$(\beta_* \mathscr{F})_x = \mathrm{ind.lim}_{U \ni x} \mathscr{F}_1(\beta^{-1}(U))$$

and $(\beta_* \mathscr{F})(U) = \mathscr{F}(\beta^{-1}(U))$.

If $A \subset X$, the sheaf $\coprod_{x \in A} \mathscr{F}_x$ is denoted by $\mathscr{F}|_A$. Here $\mathscr{F}|_A(U) :=$ ind.lim$_{V \supset U} \mathscr{F}(V)$. $\mathscr{F}|_A$ is called the **restriction** of $\mathscr{F}$ to A. A is called the **support** of $\mathscr{F}$ if "$\mathscr{F}_x = \{0_x\} \iff x \notin A$". The support of $\mathscr{F}$ is denoted by supp.$\mathscr{F}$.

Sheaves of rings and sheaves of modules are defined similarly.

Definition 1.11. A **ringed space** is a topological space equipped with a sheaf of rings.

For any complex manifold M, the family $\{\mathscr{O}(U); U$ is open in $M\}$ is naturally regarded as a sheaf by identifying an element of $\mathscr{O}(U)$ as the collection of its germs. This sheaf is called the **structure sheaf** of M and denoted by $\mathscr{O}_M$, or simply by $\mathscr{O}$. $(M, \mathscr{O})$ is the most important example of ringed space for our purpose. We note that the domains of holomorphy are nothing but the connected components of $\mathscr{O}_{\mathbb{C}^n}$. For meromorphic functions, domains of meromorphy can be characterized similarly. Namely, in the sheaf theoretic terms, meromorphic functions are identified as the sections of a sheaf in the following way: Let $\mathscr{M}_x$ be the quotient field of $\mathscr{O}_x$, let

$$\mathscr{M} = \coprod_{x \in M} \mathscr{M}_x,$$

and

$$\mathscr{M}(U) = \{h \in \mathscr{M}[U]; \text{ for every } x \in U \text{ there exist a neighborhood } V \ni x \text{ and}$$

$$f, g \in \mathscr{O}(V) \text{ such that } h(y) = \frac{f_y}{g_y} \text{ for all } y \in V\}$$

for any open set $U \subset M$. Sections of the sheaf $\{\mathscr{M}(U)\}_U$ are called **meromorphic functions**. Connected components of the sheaf $\mathscr{M}$ as the topological space are called the **domains of meromorphy**.

A sheaf $p_1 : \mathscr{F}_1 \to X_1$ is said to be **isomorphic** to a sheaf $p_2 : \mathscr{F}_2 \to X_2$ if there exists a homeomorphism $\psi : X_1 \to X_2$ and a bijection $\beta : \mathscr{F}_1 \to \mathscr{F}_2$ such that

$$\beta|_{\mathscr{F}_{1,x}} \in \text{Hom}(\mathscr{F}_{1,x}, \mathscr{F}_{2,\psi(x)})$$

for all $x \in X_1$.

Complex manifolds are naturally identified with ringed spaces which are locally isomorphic to $(D, \mathscr{O}_D)$ for some domain D in $\mathbb{C}^n$.

Definition 1.12. An ideal sheaf of a ringed space $(X, \mathscr{R})$ is a sheaf of $\mathscr{R}$-modules $(X, \mathscr{I})$ such that $\mathscr{I}_x$ is an ideal of $\mathscr{R}_x$ for each $x \in X$.

Let $\mathscr{R} \to X$ be a sheaf of commutative rings with units and let $\mathscr{E}_j \to X$ ($j = 1, 2$) be sheaves of $\mathscr{R}$-**modules** (i.e. $\mathscr{E}_{j,x}$ are $\mathscr{R}_x$-modules, etc.). A collection of $\mathscr{R}_x$-homomorphisms

$$\alpha_x : \mathscr{E}_{1,x} \longrightarrow \mathscr{E}_{2,x}, \quad x \in X,$$

denoted by $\alpha : \mathscr{E}_1 \to \mathscr{E}_2$ is called a **homomorphism** between $\mathscr{R}$-modules if

$$s \in \mathscr{E}_1(U) \Rightarrow \alpha \circ s \in \mathscr{E}_2(U)$$

holds for any open set $U \subset X$. $\alpha \circ s$ will also be denoted by $\alpha(s)$ (for a typographical reason). For the sheaves of Abelian groups and those of rings, homomorphisms are defined similarly. Sheaves of $\mathscr{O}$-modules are called **analytic sheaves**.

The stalkwise direct sum $\mathscr{R}^{\oplus m}$ is called a **free $\mathscr{R}$-module of rank** m. A sheaf of $\mathscr{R}$-modules is called **locally free** if it is locally isomorphic to a free sheaf. Locally free sheaves of rank one are said to be **invertible**. A holomorphic map ψ between two complex manifolds $(M_j, \mathscr{O}_j)$ $(j = 1, 2)$ induces a homomorphism

$$\psi_* : \mathscr{O}_2|_{\psi(M_1)} \longrightarrow \psi_* \mathscr{O}_1|_{\psi(M_1)}$$

by $\psi_*(f_{\psi(x)}) = (f \circ \psi)_{\psi(x)}$. Conversely, a continuous map $\psi : M_1 \to M_2$ is holomorphic if there exists a homomorphism $\beta : \mathscr{O}_2|_{\psi(M_1)} \to \psi_* \mathscr{O}_1|_{\psi(M_1)}$ which induces at every point $x \in M_1$ a homomorphism from $\mathscr{O}_{2,\psi(x)}$ to $\mathscr{O}_{1,x}$ which maps the invertible elements of $\mathscr{O}_{2,\psi(x)}$ to those of $\mathscr{O}_{1,x}$.

For any homomorphism $\alpha : \mathscr{E}_1 \to \mathscr{E}_2$, the collection of preimages of 0, which is called the **kernel** of α, is naturally equipped with a sheaf structure whose sections over U are precisely the elements of $\{s \in \mathscr{E}_1(U); \alpha \circ s = 0\}$, the kernel of the homomorphism

$$\alpha_U : \mathscr{E}_1(U) \ni s \to \alpha \circ s \in \mathscr{E}_2(U).$$

The kernel of α will be denoted by $\mathrm{Ker}\alpha$. Definition of the cokernel of α is more delicate: Let

$$\mathrm{coker}\alpha = \coprod_{x \in X} \mathrm{coker}\alpha_x,$$

let $\pi : \mathscr{E}_2 \to \coprod_{x \in X} \mathrm{coker}\alpha_x$ be the canonical projection, and let

$$\mathrm{coker}\alpha(U) = \{s \in \mathrm{coker}\alpha[U]; s = \pi \circ \hat{s} \text{ for some } \hat{s} \in \mathscr{E}_2(U)\}.$$

Then $\{\mathrm{coker}\alpha(U)\}$ is clearly a presheaf. The sheafification of $\{\mathrm{coker}\alpha(U)\}$ will be called the **cokernel sheaf** of α and denoted by $\mathrm{Coker}\alpha$. When α is an inclusion, $\mathrm{Coker}\alpha$ will be denoted by $\mathscr{E}_1/\mathscr{E}_2$. The **image sheaf** $\mathrm{Im}\alpha$ of α is defined similarly. Given an ideal sheaf $\mathscr{I}$ of $\mathscr{R}$, the cokernel $\mathscr{R}/\mathscr{I}$ of the inclusion morphism $\iota : \mathscr{I} \to \mathscr{R}$ carries naturally the induced structure of a sheaf of commutative rings.

A sequence

$$\cdots \to \mathscr{E}^k \to \mathscr{E}^{k+1} \to \mathscr{E}^{k+2} \to \cdots \tag{1.4}$$

of sheaves of Abelian groups or $\mathcal{R}$-modules is called an **exact sequence** if, for any two successive morphisms $\alpha^k : \mathcal{E}^k \to \mathcal{E}^{k+1}$ and $\alpha^{k+1} : \mathcal{E}^{k+1} \to \mathcal{E}^{k+2}$, $\mathrm{Im}\alpha^k = \mathrm{Ker}\alpha^{k+1}$ holds. The family $\mathcal{E}^* = \{(\mathcal{E}^k, \alpha^k)\}$ is called a **complex** of sheaves if $\mathrm{Im}\alpha^k \subset \mathrm{Ker}\alpha^{k+1}$ holds for all k. A **resolution** of a sheaf $\mathcal{F}$ is by definition an exact sequence of the form

$$0 \to \mathcal{F} \to \mathcal{E}^0 \to \mathcal{E}^1 \to \cdots .$$

Definition 1.13. The **canonical flabby resolution** of a sheaf $\mathcal{F} \to X$ is a complex $\mathcal{F}^* = \{(\mathcal{F}^k, j^k)\}_{k \in \mathbb{Z}}$ defined by

$$\mathcal{F}^k = 0 \ (= \coprod \{0_x\}) \ for \ k \leq -1,$$

$$\mathcal{F}^0 = \hat{\mathcal{F}}, \ j^{-1} = 0,$$

$$\mathcal{F}^{k+1} = (\mathrm{Coker}j^k)^\wedge \qquad (the\ canonical\ flabby\ extension)$$

and $j^{k+1} =$ the composite of the canonical projection $\mathcal{F}^k \to \mathrm{Coker}j^k$ and the inclusion $\mathrm{Coker}j^k \hookrightarrow (\mathrm{Coker}j^k)^\wedge$, for $k \geq 0$, inductively.

Clearly, $\{(\mathcal{F}^k, j^k)\}$ is a complex of sheaves and the sequence

$$0 \to \mathcal{F} \to \mathcal{F}^0 \to \mathcal{F}^1 \to \mathcal{F}^2 \to \cdots \tag{1.5}$$

is exact.

Definition 1.14. The **p-th cohomology group** of X with values in the sheaf $\mathcal{F} \to X$, denoted by $H^p(X, \mathcal{F})$, is by definition the p-th cohomology group of the complex $\{(\mathcal{F}^k(X), j_X^k)\}$.

The elements of $H^p(X, \mathcal{F})$ will be referred to as the **$\mathcal{F}$-valued p-th cohomology classes**. The restriction homomorphism $\mathcal{F}(U) \to \mathcal{F}(V)$ naturally induces a homomorphism $H^p(U, \mathcal{F}) \to H^p(V, \mathcal{F})$.

Note that $H^0(X, \mathcal{F}) = \mathcal{F}(X)$. As for $H^p(V, \mathcal{F})$, $p \geq 1$, let us briefly recall a description of the cohomology classes in $H^1(X, \mathcal{F})$. Given any $v \in \mathrm{Ker}j_X^1$, there exists an open covering $\mathcal{U} = \{U_\ell\}$ of X and $u_\ell \in \mathcal{F}^1(U_\ell)$ such that $j^1(u_\ell) = v$ holds on U_ℓ. Hence $u_\ell - u_{\ell'} \in \mathrm{Ker}j^1_{U_\ell \cap U_{\ell'}} = \mathcal{F}(U_\ell \cap U_{\ell'})$. If the cohomology class represented by v is zero, there exists $u \in \mathcal{F}^0$ such that $j^0(u) = v$. As a result one has $u_\ell - u \in \mathcal{F}(U_\ell)$. Therefore the collection of $u_\ell - u_{\ell'}$, as an element of $\oplus_{\ell,\ell'} \mathrm{Ker}j^1_{U_\ell \cap U_{\ell'}}$, is in the image of the map

$$\delta^0 : \oplus_\ell \mathcal{F}(U_\ell) \ni \{u_\ell\}_\ell \mapsto \{u_\ell - u_{\ell'}\}_{\ell,\ell'} \in \oplus_{\ell,\ell'} \mathcal{F}(U_\ell \cap U_{\ell'}).$$

Consequently, letting

$$C^p(\mathcal{U}, \mathcal{F}) = \oplus_{\ell_0,\ldots,\ell_p} \{u_{\ell_0\ldots\ell_p} \in \mathcal{F}(U_{\ell_0} \cap \cdots \cap U_{\ell_p}); u_{\ell_0\ldots\ell_p} \ is \ alternating$$

$$in \ \ell_0, \ldots, \ell_p\}$$

and defining

$$\delta^p_{\mathscr{U}} : C^p(\mathscr{U},\mathscr{F}) \longrightarrow C^{p+1}(\mathscr{U},\mathscr{F})$$

and $H^p(\mathscr{U},\mathscr{F})$ respectively by

$$\delta^p_{\mathscr{U}}(\{u_{\ell_0\ldots\ell_p}\}_{\ell_0,\ldots,\ell_p}) = \left\{ \sum_{0 \le j \le p} (-1)^j u_{\ell'_0\ldots\ell'_{j-1}\ell'_{j+1}\ldots\ell'_{p+1}} \right\}_{\ell'_0,\ldots,\ell'_{p+1}}$$

and $\mathrm{Ker}\,\delta^p_{\mathscr{U}}/\mathrm{Im}\,\delta^{p-1}_{\mathscr{U}}$, one has a homomorphism

$$\gamma^1 : H^1(X,\mathscr{F}) \to \mathrm{ind.lim}_{\mathscr{U}} H^1(\mathscr{U},\mathscr{F})$$

defined by the correspondence $[v] \to [\{u_\ell - u'_\ell\}_{\ell,\ell'}]$, and similarly $\gamma^p : H^p(X,\mathscr{F}) \to$ ind.lim$_{\mathscr{U}} H^p(\mathscr{U},\mathscr{F})$ for all p. Here the inductive system $\{H^p(\mathscr{U},\mathscr{F})\}$ is with respect to the restriction homomorphisms $H^p(\mathscr{U},\mathscr{F}) \to H^p(\mathscr{V},\mathscr{F})$ for the refinements $\mathscr{V}$ of $\mathscr{U}$. See [G-R] (for instance) for the detail of the construction of γ^p for $p \ge 1$ and for the proof of the following extremely important fact.

Theorem 1.18. *γ^p are isomorphisms if X is a paracompact Hausdorff space.*

For any paracompact Hausdorff space X, a sheaf $\mathscr{G} \to X$ of Abelian groups is said to be **fine** if, given any open covering $\mathscr{U}$ of X, there exists a locally finite refinement $\mathscr{V} = \{V_j\}$ of $\mathscr{U}$ and homomorphisms $h_j : \mathscr{G} \to \mathscr{G}$ such that

$$\mathrm{supp}\,h_j := \overline{\{x \,;\, h_j|\mathscr{G}_x \ne 0\}} \subset V_j$$

and $\sum_j h_j = 1$.

Corollary of Theorem 1.18. If $\mathscr{G} \to X$ is a fine sheaf,

$$H^p(X,\mathscr{G}) = 0$$

for any $p \ge 1$.

Another basic fact is the existence of a canonically defined exact sequences of the cohomology groups: Let

$$0 \to \mathscr{E} \to \mathscr{F} \to \mathscr{G} \to 0 \tag{1.6}$$

be an exact sequence of sheaves over X. Then it is easy to see that the induced sequence

$$0 \to \mathscr{E}(X) \to \mathscr{F}(X) \to \mathscr{G}(X)$$

is exact. By the exactness of (1.6), this sequence can be prolonged canonically as

$$\mathscr{E}(X) \to \mathscr{F}(X) \to \mathscr{G}(X) \to H^1(X,\mathscr{E}) \to H^1(X,\mathscr{F}) \to H^1(X,\mathscr{G}) \to H^2(X,\mathscr{E}) \to \cdots,$$

which is called the **long exact sequence** associated to (1.6) (cf. [G-R]).

1.3.2   Coherent Sheaves, Complex Spaces, and Theorems A and B

In the study of ideals of holomorphic functions, Oka and Cartan were led to introduce a notion characterizing a class of ideal sheaves of $\mathscr{O}$, the *coherence* (cf. [O-2] and [C]).

Definition 1.15. An $\mathscr{R}$-module $\mathscr{E}$ over a topological space X is called **coherent** if:

(1) $\mathscr{E}$ is **locally finitely generated**, i.e., for any $x_0 \in X$ there exist a neighborhood $U \ni x_0$ and finitely many sections of $\mathscr{M}$ over U whose values at $x \in U$ generate $\mathscr{E}_x$ over $\mathscr{R}_x$ for any $x \in U$.
(2) For any $m \in \mathbb{N}$ and for any morphism α from the direct sum $\mathscr{R}^{\oplus m}$ to $\mathscr{E}$, $\mathrm{Ker}\,\alpha$ is locally finitely generated.

A penetrating insight (definitely shared by Oka and Cartan) was that a principal basic question of several complex variables is to establish a criterion for the analytic sheaves to be globally generated.

For any complex manifold $(M, \mathscr{O})$, Oka established the following basic result by exploiting the Weierstrass division theorem to run an induction argument on the dimension.

Theorem 1.19 (Oka's coherence theorem). $\mathscr{O}$ *is coherent.*

For the proof, the reader is referred to [G-R, Hö-2], or [Nog]. By this theorem, for any coherent $\mathscr{O}$-module $\mathscr{F}$ and for any $x \in M$, one can find a neighborhood $U \ni x$ and an exact sequence over U of the form

$$\cdots \to \mathscr{O}^{\oplus m_k}|_U \to \cdots \to \mathscr{O}^{\oplus m_2}|_U \to \mathscr{O}^{\oplus m_1}|_U \to \mathscr{F}|_U \to 0,$$

which is called a **free resolution** of $\mathscr{F}$ over U. Since $\mathbb{C}\{z_1,\ldots,z_n\}$ is a regular local ring of dimension n, the kernel of $\mathscr{O}^{\oplus m_n}|_U \to \mathscr{O}^{\oplus m_{n-1}}|_U$ is locally free by Hilbert's syzygy theorem (cf. [G-R]).

For any $A \subset M$, $\mathscr{I}_A$ will stand for the ideal sheaf of $\mathscr{O}$ consisting of the germs of holomorphic functions vanishing along A, i.e.

$$\mathscr{I}_A(U) = \{f \in \mathscr{O}(U); f|_{U \cap A} = 0\}.$$

$\mathscr{I}_A$ is called the **ideal sheaf of A**, for short.

Definition 1.16. A closed set $A \subset M$ is called an **analytic set** if for every point $x \in A$ there exist a neighborhood U, $m \in \mathbb{N}$, and $f_1, \ldots, f_m \in \mathcal{O}(U)$ such that $U \cap A = \{w \in U; f_1(w) = \ldots = f_m(w) = 0\}$.

From the definition, it is clear that $\mathrm{supp}(\mathcal{O}/\mathcal{I})$ is analytic if $\mathcal{I}$ is a coherent ideal sheaf. The vector–valued holomorphic function $(f_1, \ldots, f_m)$ is called a **local defining function** of A around x. By the **codimension** of an analytic set A at $x \in A$, we shall mean the minimal number of holomorphic functions $f_1, \ldots, f_k$ defined on a neighborhood of U such that x is isolated in $A \cap \left(\cap_{j=1}^{k} f_j^{-1}(0) \right)$.

Theorem 1.20 (Rückert's Nullstellensatz). *Let $(f_1, \ldots, f_m)$ be a local defining function of an analytic set A around x and let $f \in \mathcal{I}_{A,x}$. Then there exists $p \in \mathbb{N}$ such that*

$$f^p \in \sum_{j=1}^{m} f_j \cdot \mathcal{O}_{M,x}.$$

For the proof, the reader is referred to [G-R, Chapter 3, A].

Theorem 1.21 (Cartan's coherence theorem). $\mathcal{I}_A$ *is coherent if A is analytic.*

Sketch of Proof. Let $x \in A$, let $(f_1, \ldots, f_m)$ be a local defining function of A around x, and let $\mathcal{I}_A$ be the ideal sheaf generated by f_j $(1 \leq j \leq m)$ over a neighborhood $U \ni x$. Then one has an exact sequence of $\mathcal{O}$-modules

$$0 \to \hat{\mathcal{I}}_A \to \mathcal{I}_A|_U \to \mathcal{I}_A|_U / \hat{\mathcal{I}}_A \to 0.$$

Since $\hat{\mathcal{I}}_A$ is coherent by Theorem 1.19, the coherence of $\mathcal{I}_A$ follows by a descending induction on the codimension of A. □

Remark 1.6. The ideal sheaf $\mathcal{I}$ of the form

$$\mathcal{I}_x = \{f_x \in \mathcal{O}_x; \int_U |f|^2 e^{-\varphi} d\lambda < \infty \text{ for some neighborhood } U \text{ of } x\}$$

turns out to be coherent if φ is plurisubharmonic (see Chap. 3).

Roughly speaking, complex spaces are complex manifolds with singularities. In functions theory, such things arise as ringed spaces which are locally isomorphic to those whose underlying spaces are the sets of common zeros of holomorphic functions.

Definition 1.17. A ringed space $(X, \mathcal{O})$ is called a **complex space** if every point $x \in X$ has a neighborhood U such that $(U, \mathcal{O}|_U)$ is isomorphic to $(\mathrm{supp}(\mathcal{O}_D/\mathcal{I}), \mathcal{O}_D/\mathcal{I})$ for some domain D in $\mathbb{C}^N$ $(N = N(x))$ and for some coherent ideal sheaf $\mathcal{I}$ of $\mathcal{O}_D$.

X will be referred to as the **underlying space** of the complex space $(X, \mathcal{O})$. $(X, \mathcal{O})$ is said to be compact if so is the underlying space X. A point $x \in X$ is called a **regular point** of $(X, \mathscr{O})$ if one can find U and D such that $\mathscr{O}|_U \cong \mathscr{O}_D$. The set of regular points of X is denoted by X_{reg}. $X \setminus X_{reg}$ is denoted by SingX. $\mathscr{O}$ is called the **structure sheaf** of X. The structure sheaf is denoted also by $\mathscr{O}_X$. Coherent $\mathscr{O}$-sheaves will be called **coherent analytic sheaves**. A closed set $A \subset X$ is called an **analytic set** if it is the support of some coherent analytic sheaf over X. An analytic set of X is naturally equipped with the structure of a reduced complex space induced from $\mathscr{O}_X$. Analytic sets of $\mathbb{CP}^n$ are called **projective algebraic sets**. The implicit function theorem naturally implies that SingX is an analytic set of X. X is called **nonsingular** if SingX $= \varnothing$. $(X, \mathscr{O})$ is said to be **irreducible** if every proper analytic set is nowhere dense. Irreducible complex spaces are called **varieties**.

By a routine argument one can infer the following from Theorem 1.19.

Theorem 1.22. *The structure sheaf of a complex space is coherent.*

For any complex space $(X, \mathscr{O})$, the elements of $\mathscr{O}(X)$ will be called **holomorphic functions** on X. Holomorphic functions on X naturally induce $\mathbb{C}$-valued functions on X. If a holomorphic function f on X is zero as a function, then, for each $x \in X$ f_x is nilpotent, i.e. some power of f_x is zero, by Rückert's Nullstellensatz. By an abuse of notation, the values of f in $\mathbb{C}$ will be denoted by $f(x)$.

Given two complex spaces $(X, \mathscr{O}_X)$ and $(Y, \mathscr{O}_Y)$, a **holomorphic map** from $(X, \mathscr{O}_X)$ to $(Y, \mathscr{O}_Y)$ is by definition a pair of continuous map $\psi : X \to Y$ and a homomorphism

$$\beta : \mathscr{O}_Y|_{\psi(X)} \longrightarrow \psi_* \mathscr{O}_X|_{\psi(X)}$$

of $\mathscr{O}|_{\psi(X)}$-modules which maps invertible elements to invertible elements.

For any holomorphic map (ψ, β) from $(X, \mathscr{O}_X)$ to $(Y, \mathscr{O}_Y)$, a homomorphism from $H^p(Y, \mathscr{O}_Y)$ to $H^p(X, \mathscr{O}_X)$ is induced canonically. For simplicity, (ψ, β) will be referred to as ψ and the induced homomorphism $H^p(Y, \mathscr{O}_Y) \to H^p(X, \mathscr{O}_X)$ by ψ^*.

Definition 1.18. A complex space $(X, \mathscr{O})$ is said to be **reduced** if no stalk of $\mathscr{O}$ contains a nilpotent element.

For any reduced complex space $(X, \mathscr{O})$ and $f, g \in \mathscr{O}(X)$, $f = g$ if and only if $f(x) = g(x) (\in \mathbb{C})$ for all $x \in X$. For any complex space $(X, \mathscr{O})$, the collection of the nilpotent elements in the stalks of $\mathscr{O}$ is an ideal sheaf, say $\mathscr{J}$. Then $(X, \mathscr{O}/\mathscr{J})$ is a reduced complex space. We shall call it the **reduction** of $(X, \mathscr{O})$.

If $(X, \mathscr{O})$ is reduced, X_{reg} is an everywhere dense subset of X. We then put dimX $:=$ dimX_{reg} and

$$\dim_x X = \inf \{\dim U_{reg}; U \text{ is a neighborhood of } x\} \quad \text{for any } x \in X.$$

Generally, $\dim_x X$ will stand for that of the reduction of $(X, \mathscr{O})$. X is called a **complex curve** if dimX $= 1$.

Definition 1.19. A complex space $(X, \mathcal{O})$ is called a **Stein space** if any discrete closed subset of X is mapped injectively into a discrete closed subset of $\mathbb{C}$ by a holomorphic function on X.

Theorem 1.23 (Cartan's theorem A). *Let $(X, \mathcal{O})$ be a Stein space. Then, for any coherent analytic sheaf $\mathcal{F}$ over X and for any point $x \in X$, the image of the natural restriction map*

$$\mathcal{F}(X) \longrightarrow \mathcal{F}_x$$

generates $\mathcal{F}_x$ over $\mathcal{O}_x$.

Combining Theorem 1.23 with Cartan's coherence theorem, we obtain for instance the following.

Proposition 1.1. *Let $(X, \mathcal{O})$ be a Stein space, let $A \subset M$ be an analytic set, and let $x \in A$ be any point. Then there exist a neighborhood $U \ni x$ and $f_1, \ldots, f_m \in \mathcal{O}(X)$ such that $U \cap A = \{y \in U; f_1(y) = \ldots = f_m(y) = 0\}$.*

Theorem 1.23 was first established by Oka when X is a domain of holomorphy in $\mathbb{C}^n$ and $\mathcal{F}$ is a coherent ideal sheaf (*idéal de domaine indéterminé*) over X. For the proof, Oka and Cartan solved a problem which P. Cousin had solved in 1895 in a very special case to construct meromorphic functions with given poles on the products of plane domains. This argument was extended eventually to show that Theorem 1.23 is a consequence of the assertion that $H^1(X, \mathcal{F}) = 0$ holds for any coherent analytic sheaf over a Stein space $(X, \mathcal{O})$. An ultimately strengthened form of such a cohomology vanishing theorem on Stein spaces is the following.

Theorem 1.24 (Cartan's theorem B). $H^p(X, \mathcal{F}) = 0$ *for any $p \geq 1$ if $\mathcal{F}$ is a coherent analytic sheaf over a Stein space $(X, \mathcal{O})$.*

There are a lot of implications of the vanishing of cohomology groups. It may be worthwhile to recall that there is a characterization of Stein spaces among them.

Theorem 1.25. $(X, \mathcal{O})$ *is a Stein space if and only if $H^1(X, \mathcal{I}) = 0$ for any coherent ideal sheaf of $\mathcal{O}$.*

Proof. For any discrete closed set $\Gamma \subset X$, the ideal sheaf $\mathcal{I}_\Gamma$ of Γ is coherent by Theorem 1.21. Hence $H^1(X, \mathcal{I}_\Gamma) = 0$, so that from the exact sequence

$$0 \to \mathcal{I}_\Gamma \to \mathcal{O}_X \to \mathcal{O}_X/\mathcal{I}_\Gamma \to 0$$

one has the surjectivity of the natural restriction homomorphism $\mathcal{O}(X) \to \mathbb{C}^\Gamma$. $\square$

Remark 1.7. The surjectivity of $\mathcal{O}(X) \to \mathbb{C}^\Gamma$ is equivalent to the injectivity of $H^1(X, \mathcal{I}) \to H^1(X, \mathcal{O})$ by the long exact sequence. This point has a significance in the development of the application of the L^2 technique. If $(X, \mathcal{O})$ is a domain over $\mathbb{C}^n$, it is easy to see that X is Stein if and only if $H^q(X, \mathcal{O}) = 0$ for any $1 \leq q \leq n-1$

(cf. [Oh-21, 2.3]). Actually, $H^n(X, \mathcal{O}) = 0$ holds whenever X is a complex space of dimension n which does not contain any compact n-dimensional analytic subset (cf. [Siu-1]).

Once it is recognized that Theorem 1.23 is a consequence of a vanishing theorem as above, it is not difficult to show the following for instance.

Theorem 1.26 (cf. [Gra-2]). *Every analytic subset of a Stein space of dimension n is the set of common zeros of at most n + 1 holomorphic functions.*

Remark 1.8. It was known by Kronecker in the nineteenth century that every algebraic set in $\mathbb{C}^n$ is the set of common zeros of $n + 1$ polynomials. Eisenbud and Evans proved in [Eb-E] that n polynomials suffice.

The L^2 method provides another effective way to analyze the sheaf cohomology groups. As a result, Theorem 1.24 can be extended to pseudoconvex spaces and certain ideal sheaves arising naturally in basic questions of complex geometry. (See Chap. 3.)

1.3.3 Coherence of Direct Images and a Theorem of Andreotti and Grauert

A complex space $(X, \mathcal{O})$ is said to be **normal** if, for any open set $U \subset X$ and for any nowhere–dense analytic set $A \subset U$, bounded holomorphic functions on $U \setminus A$ extend holomorphically to U. In [O-3], Oka proved that every reduced complex space has a "normal model", i.e., for any reduced complex space $(X, \mathcal{O})$ there exist a reduced complex space $(\hat{X}, \mathcal{O}_{\hat{X}})$ and a proper surjective holomorphic map $p : \hat{X} \to X$ such that $p|_{p^{-1}(X_{reg})}$ is a biholomorphic map onto X_{reg} and the inclusion map $\iota : \hat{X}_{reg} \hookrightarrow \hat{X}$ induces an isomorphism $\iota_* \mathcal{O} \cong \mathcal{O}_{\hat{X}}$. It can be stated as another coherence theorem as follows.

Theorem 1.27 (Oka's normalization theorem). *Let $(X, \mathcal{O})$ be a reduced complex space and let $\iota : X_{reg} \to X$ be the inclusion map. Then the direct image sheaf $\iota_* \mathcal{O}_{X_{reg}}$ is a coherent sheaf of rings if $\dim_x \mathrm{Sing} X + 2 \le \liminf_{y \to x} \dim_y X$ for all $x \in \mathrm{Sing} X$.*

Substantially, this is a primitive form of Hironaka's desingularization theorem. Recall that Hironaka's desingularization theorem asserts that every reduced complex space has a nonsingular model, i.e., for any reduced complex space $(X, \mathcal{O}_X)$ one can find a complex manifold $(M, \mathcal{O}_M)$ and a proper holomorphic map $\tilde{\pi} : M \to X$ such that $\tilde{\pi}|_{M \setminus \tilde{\pi}^{-1}(\mathrm{Sing} X)}$ is a biholomorphic map onto X_{reg}.

Let $\psi : X \to Y$ be a holomorphic map. For any analytic sheaf $\mathscr{F} \to X$, the p-**th direct image sheaf** of $\mathscr{F}$, denoted by $R^p \psi_* \mathscr{F}$, is defined as the sheafification of the presheaf

$$U \longrightarrow \{s : U \to \coprod_{y \ni U} (\text{ind.lim}_{V \ni y} H^p(\psi^{-1}(V), \mathscr{F})); \text{ there exists } u \in H^p(\psi^{-1}(U), \mathscr{F})$$

$$\text{such that } s(y) = u_y \text{ for all } y\}.$$

The following is a very profound result.

Theorem 1.28 (cf. [Gra-4, Gra-R-2]). *Let $\psi : X \to Y$ be a proper holomorphic map between complex spaces and let $\mathscr{F} \to X$ be a coherent analytic sheaf. Then $R^p \psi_* \mathscr{F}$ $(p \geq 0)$ are coherent analytic sheaves over Y.*

Corollary 1.2 (Remmert's proper mapping theorem). *For any proper holomorphic map $\psi : X \to Y$, $\psi(X)$ is an analytic set of Y.*

Theorem 1.28 is a generalization of the following.

Theorem 1.29 (Cartan–Serre finiteness theorem). *For any compact complex space $(X, \mathscr{O})$ and for any coherent analytic sheaf $\mathscr{F}$ over X, $\dim H^p(X, \mathscr{F}) < \infty$ for all p.*

Andreotti and Grauert studied intermediate results between Theorems 1.24 and 1.29. For that they introduced a class of q-convex spaces.

Definition 1.20. A continuous function $\phi : X \to \mathbb{R}$ is called q-**convex** at $x \in X$ if one can find a neighborhood $U \ni x$, a domain $D \supset \mathbb{C}^N$, a coherent ideal sheaf $\mathscr{I} \subset \mathscr{O}_D$ such that

$$(U, \mathscr{O}_U) \cong (\text{supp}(\mathscr{O}_D/\mathscr{I}), \mathscr{O}_D/\mathscr{I}) \tag{1.7}$$

and a q-convex function $\tilde{\phi}_D$ on D whose restriction to $\text{supp}(\mathscr{O}_D/\mathscr{I})$ coincides with $\phi|U$ by the isomorphism (1.7).

Definition 1.21. A complex space $(X, \mathscr{O})$ is said to be q-**complete** (resp. q-**convex**) if X admits a continuous exhaustion function which is q-convex everywhere (resp. q-convex outside a compact subset of X).

In view of Theorem 1.15, the following is a generalization of Theorem 1.24.

Theorem 1.30 (cf. Andreotti and Grauert [A-G]). *Let X be a q-complete (resp. q-convex) space and let $\mathscr{F} \to X$ be a coherent analytic sheaf. Then $H^p(X, \mathscr{F})$ are 0 (resp. finite dimensional) for $p \geq q$.*

However, it is not known, except for the cases $q = 1$ and $q \geq \dim X$, whether or not $(X, \mathscr{O})$ is q-complete (resp. q-convex) if $H^p(X, \mathscr{F})$ are all zero (resp. finite dimensional) for any $p \geq q$ and for any analytic sheaf $\mathscr{F}$. The so-called Grauert conjecture is "It is the case". A reason to expect it is in the theory of cycle spaces. It is known that the set of certain equivalence classes of holomorphic maps from compact complex manifolds to a complex space X is canonically equipped with a structure of a complex space, say $\mathscr{B}(X)$ (cf. [B-1]), and the connected components of $\mathscr{B}(X)$ for the maps from purely $(q-1)$-dimensional manifolds are Stein if X is q-complete (cf. [B-2]). (See [G-W-1, Dm-1] and [Oh-9] for the case $q = \dim X$.) For q-convex manifolds M and locally free analytic sheaves $\mathscr{F}$ over M, one can analyze $H^p(M, \mathscr{F})$ by the L^2 method. For instance, one has an extension of the Hodge theory on compact complex manifolds to certain q-convex manifolds (cf. Chap. 2).

1.4 $\bar{\partial}$-Equations on Manifolds

By extending Theorem 1.18, one has a very basic result on the cohomology groups of complex manifolds with values in the locally free analytic sheaves, the Dolbeault isomorphism theorem. Based on this, one can represent the cohomology classes by differential forms. The cohomology classes in $H^p(X, \mathscr{F})$ can be studied in a way closely related to the geometry of X and $\mathscr{F}$ by exploiting such an expression. Basic notions needed for that are holomorphic vector bundles, Dolbeault cohomology groups, Chern connections, curvature forms, etc., which belong to differential geometry and recalled below together with classical results as Dolbeault's isomorphism theorem and Serre's duality theorem, which are natural counterparts of the de Rham isomorphism and the Poincaré duality in the classical global differential geometry.

1.4.1 Holomorphic Vector Bundles and $\bar{\partial}$-Cohomology

A **holomorphic vector bundle** over a complex manifold is by definition a C^∞ complex vector bundle whose transition functions are holomorphic. In other words, a complex manifold E is called a holomorphic vector bundle over a complex manifold M if E is of the form $\coprod_{x \in M} E_x$, as a set for some vector spaces E_x parametrized by M, satisfying the following requirements:

(1) The map

$$\pi : E \supset E_x \mapsto \{x\} \subset M$$

   is holomorphic and $d\pi : T_E^{1,0} \to T_M^{1,0}$ is everywhere of maximal rank.
(2) For any open set $U \subset M, f, g \in \mathscr{O}(U)$ and holomorphic sections $s, t : U \to E$, $fs + gt$ is also holomorphic.

The rank of E ($= \mathrm{rank}E := \dim \mathrm{Ker}d\pi$) is a locally constant function on M, which will be assumed to be constant unless stated otherwise. From the definition, it is clear that the direct sums, the duals and the tensor products of holomorphic vector bundles are naturally defined by fiberwise construction. They will be denoted by $E_1 \oplus E_2, E^*$ and $E_1 \otimes E_2$, respectively. $\mathrm{Hom}(E_1, E_2)$ will stand for $E_1^* \otimes E_2$. $\wedge^r E$ will be denoted by $\det E$ if $r = \mathrm{rank}E$. Holomorphic vector bundles over complex spaces are defined similarly. For any subset $A \subset M, E|_A$ will stand for the vector bundle $\pi^{-1}(A) \to A$. A **local (holomorphic) frame** of E over an open set U is an r-tuple of holomorphic sections, say $s_1, \ldots, s_r$ of $E|U \to U$ such that $s_1(x), \ldots, s_r(x)$ are linearly independent for all $x \in U$.

For two holomorphic vector bundles E_j ($j = 1, 2$) over M, a holomorphic map $\alpha : E_1 \to E_2$ is called a **bundle homomorphism**, or simply a homomorphism, if $\alpha|_{E_{1,x}} \in \mathrm{Hom}(E_{1,x}, E_{2,x})$ for all $x \in M$ and $\dim \alpha(E_{1,x})$ is a locally constant function on M. The isomorphism $E_1 \cong E_2$ will mean that there exists a biholomorphic bundle homomorphism from E_1 to E_2. Given any bundle homomorphism, its kernel and cokernel are naturally defined as holomorphic vector bundles. Subbundles and quotient bundles are defined similarly. A sequence of bundle homomorphisms is said to be **exact** if it is exact fiberwise. Given a holomorphic map $f : M \to N$ and a holomorphic vector bundle $\pi : E \to N, f^*E$ will stand for the fiber product of f and π, naturally equipped with the structure of a holommorphic vector bundle over M. Holomorphic vector bundles of rank one are called **holomorphic line bundles**. For any holomorphic line bundle $L \to M$ and $m \in \mathbb{N} \cup \{0\}, L^{\otimes m}$ will be denoted simply by L^m, since there will be no fear of confusing the abbreviated $L^{\otimes 2}$ with "square integrable". If $-m \in \mathbb{N}, L^m := (L^*)^{(-m)}$.

Important examples of holomorphic vector bundles are $T_M^{p,0}$ and $(T_M^{p,0})^*$. If M is of pure dimension n, $(T^{n,0})^*$ is called the **canonical bundle** and denoted by $\mathbb{K}_M$. Let S be a (not necessarily closed) complex submanifold of M. Then $T_M^{1,0}|_S$ is a holomorphic vector bundle over S, so that one has a natural inclusion homomorphism $T_S^{1,0} \hookrightarrow T_M^{1,0}|_S$. The quotient bundle $T_M^{1,0}|_S / T_S^{1,0}$ is by definition the holomorphic normal bundle of S in M, and denoted by $N_{S/M}$, or more explicitly by $N_{S/M}^{1,0}$ in some context. There is a neat relation between the canonical bundles of M and S:

$$\mathbb{K}_M|_S \cong \mathbb{K}_S \otimes \det N_{S/M}^{-1} \qquad \textbf{(adjunction formula)}$$

Let us mention a more specific example.

Example 1.7. By extending the natural projection

$$\pi : \mathbb{C}^{n+1} \setminus \{0\} \to \mathbb{CP}^n,$$

one has a holomorphic line bundle

$$\tilde{\pi} : \coprod_{x \in \mathbb{CP}^n} \overline{\pi^{-1}(x)} \to \mathbb{CP}^n.$$

Here $\overline{\pi^{-1}(x)}$ denotes the closure of $\pi^{-1}(x)$ in $\mathbb{C}^{n+1}$. The bundle $\coprod \overline{\pi^{-1}(x)}$ is called the **tautological line bundle** and denoted by $\tau_{\mathbb{CP}^n}$. It is easy to verify that $\mathbb{K}_{\mathbb{CP}^n} \cong \tau_{\mathbb{CP}^n}^{n+1}$. $\tau_{\mathbb{CP}^n}^{-1}$ is called the **hyperplane section bundle**.

A **local trivialization** of E around a point $x \in M$ is a chart of the form $(\pi^{-1}(U), \psi)$ such that $x \in U$ and ψ maps $\pi^{-1}(U)$ to $U \times \mathbb{C}^r$ ($r = \mathrm{rank} E$ on U)

in such a way that $pr_{\mathbb{C}^r} \circ \psi|_{E_y} \in \mathrm{Hom}(E_y, \mathbb{C}^r)$ for all $y \in U$, where $pr_{\mathbb{C}^r}$ denotes
the projection $U \times \mathbb{C}^r \to \mathbb{C}^r$. Given two local trivializations $(\pi^{-1}(U), \psi_U)$ and
$(\pi^{-1}(V), \psi_V)$, $\psi_U \circ \psi_V^{-1}(x, \zeta) = (x, e_{UV}(x) \cdot \zeta)$ holds for any $(x, \zeta) \in (U \cap V) \times \mathbb{C}^r$
for some $e_{UV} \in \mathscr{O}(U \cap V, \mathrm{GL}(r, \mathbb{C}))$. e_{UV} is called a **transition function** of E. A
system of local trivializations $\{\pi^{-1}(U), \psi_U\}_{U \in \mathscr{U}}$ associated to an open covering $\mathscr{U}$
of M yields a system of transition functions

$$\{e_{UV}; U, V \in \mathscr{U}, U \cap V \neq \varnothing\}.$$

Obviously $\{e_{UV}\}$ satisfies

$$e_{UV} e_{VW} = e_{UW} \text{ on } U \cap V \cap W \neq \varnothing. \tag{1.8}$$

Conversely, given any open covering $\mathscr{U}$ of M and a system of $\mathrm{GL}(r, \mathbb{C})$-valued
holomorphic functions e_{UV} $(U, V \in \mathscr{U})$ satisfying the transition relations (1.8), a
holomorphic vector bundle $E \to M$ is defined by

$$E = \left(\coprod_{U \in \mathscr{U}} U \times \mathbb{C}^r \right) / \sim.$$

Here the equivalence relation $\sim$ is defined by

$$U \times \mathbb{C}^r \ni (x, \zeta) \sim (y, \xi) \in V \times \mathbb{C}^r \iff x = y \text{ and } \zeta = e_{UV}(x)\xi.$$

Given a holomorphic vector bundle $E \to M$ and an open set $U \subset M$, we put

$$\Gamma(U, E) = \{s \in \mathscr{O}(U, E); \pi \circ s = id_U\},$$

$$\mathscr{O}_x^E = \bigcup_{U \ni x} \{s_x; s \in \Gamma(U, E)\} \quad (x \in M),$$

where s_x denotes the germ of s at x, and

$$\mathscr{O}^E(U) = \{\sigma \in \mathscr{O}^E[U]; \text{ there exists } s \in \Gamma(U, E) \text{ such that } s_x = \sigma(x) \text{ for all } x \in U \}.$$

Namely, we consider the sheaf $\mathscr{O}^E = \coprod_{x \in M} \mathscr{O}_x^E$ (or $\{\mathscr{O}^E(U)\}_U$). Clearly, $\mathscr{O}^E$ is
a locally free analytic sheaf. $\mathscr{O}^E$ will be identified with E in many contexts. For
simplicity, we shall denote $\mathscr{O}^{E \otimes (T_M^{p,0})^*}$ by $\Omega_{M,E}^p$. Further, by an abuse of notation,
$\Omega_{M,E}^p$ will be denoted by $\Omega^p(E)$ and $\Omega_{M,E}^0$ by $\mathscr{O}(E)$. $\mathscr{O}(E)$ is an invertible sheaf if
and only if $\mathrm{rank} E = 1$. Let $A \subset M$ be an analytic set whose ideal sheaf is invertible.
Then, there exists a holomorphic line bundle $L \to M$ such that $\mathscr{I}_A \cong \mathscr{O}(L)$. It's
a convention that $L^{-1}(= L^*)$ is denoted by $[A]$. A **divisor** on M is by definition
a formal linear combination $\sum_{j=1}^k m_j A_j$ $(m \in \mathbb{Z})$ such that $\mathscr{I}_{A_j}$ are invertible. The

bundle $\prod_{j=1}^k [A_j]^{m_j}$ is denoted additively by $\sum_{j=1}^k m_j[A_j]$ in some places. A divisor $\sum_{j=1}^k m_j A_j$ is called an **effective divisor** if $m_j > 0$ for all j. Given an effective divisor $\delta = \sum_{j=1}^k m_j A_j$, we put

$$|\delta| = \bigcup_j A_j.$$

$|\delta|$ is called the support of δ. If $M = \mathbb{CP}^n$ and A is a complex hyperplane, $[mA]$ is denoted by $\mathcal{O}(m)$ in many places. $\mathcal{O}(1)$ is called the **hyperplane section bundle**.

Similarly to the case of holomorphic vector bundles, for any C^∞ complex vector bundle $E_1 \to M$, we define the sheaf $\mathscr{C}_{M,E_1}^\infty = \coprod_{x \in M} \mathscr{C}_{M,E_1,x}^\infty$ of the germs $\mathscr{C}_{M,E_1,x}^\infty$ of C^∞ sections of E_1. $\mathscr{C}_{M,E_1}^\infty(U)$ will be denoted simply by $\mathscr{C}^\infty(U,E_1)$. If $E_1 = E_0 \otimes (T_M^{p,q})^*$ (resp. $E_1 = E_0 \otimes \wedge^r(T_M^{\mathbb{C}})^*$) for some C^∞ vector bundle E_0, we shall denote $\mathscr{C}_{M,E_1}^\infty$ simply by $\mathscr{C}^{p,q}(E_0)$ (resp. $\mathscr{C}^r(E_0)$) and $\mathscr{C}^{p,q}(E_0)(U)$ (resp. $\mathscr{C}^r(E_0)(U)$) by $C^{p,q}(U,E_0)$ (resp. $C^r(U,E_0)$). Elements of $C^{p,q}(U,E_0)$ (resp. $C^r(U,E_0)$) will be referred to as E_0-valued (p,q)-forms (resp. r-forms) on U.

For any holomorphic vector bundle $E \to M$ and for any open covering $\mathscr{U} = \{U_j\}$ of M such that $E|_{U_j} \cong U_j \times \mathbb{C}^r$, the elements of $C^{p,q}(M,E)$ are naturally identified with the systems of vector–valued C^∞ (p,q)-forms $\{u_j\}$, u_j being defined on U_j, such that $u_j = e_{jk} u_k$ ($e_{jk} := e_{U_j U_k}$) holds whenever $U_j \cap U_k \neq \varnothing$. In particular, the complex exterior derivative $\bar{\partial}$ of type $(0,1)$ maps $C^{p,q}(U,E)$ to $C^{p,q+1}(U,E)$ ($U \subset M$) so that $\bar{\partial}$ induces a complex $\{\mathscr{C}^{p,q}(E), \bar{\partial}\}_{q \geq 0}$.

The associated sequence

$$0 \to \Omega^p(E) \hookrightarrow \mathscr{C}^{p,0}(E) \to \mathscr{C}^{p,1}(E) \to \cdots \tag{1.9}$$

is exact (**Dolbeault's lemma**).

The proof of Dolbeault's lemma is based on the characterization of holomorphic functions as C^1 solutions of the Cauchy–Riemann equation and **Pompeiu's formula**

$$u(z) = \frac{1}{2\pi i} \left\{ \int_{\partial D} \frac{u(\zeta)}{\zeta - z} d\zeta + \int_D \frac{\partial u/\partial \bar{\zeta}}{\zeta - z} d\zeta \wedge d\bar{\zeta} \right\}, \quad z \in D, \tag{1.10}$$

which holds for any C^1 function u on the closure of a bounded domain $D \subset \mathbb{C}$ with smooth boundary (cf. [G-R, Hö-2]). Here

$$\int \cdot \, d\zeta \wedge d\bar{\zeta} := -2i \int \cdot \, d\lambda_1.$$

In fact, (1.10) implies in particular that, for any compactly supported C^∞ function φ on $\mathbb{C}$, the function

$$u(z) = \frac{-1}{2\pi i} \int_{\mathbb{C}} \frac{\varphi(z - \zeta)}{\zeta} d\zeta \wedge d\bar{\zeta}$$

satisfies the equation

$$\frac{\partial u}{\partial \bar{z}} = \varphi,$$

so that an induction argument works to prove the exactness of (1.9) (for the detail, see [G-R] or [W] for instance). The sequence (1.9) is called the **Dolbeault complex**.

Definition 1.22.

$$H^{p,q}(M,E) := \operatorname{Ker}\bar{\partial} \cap C^{p,q}(M,E)/\operatorname{Im}\bar{\partial} \cap C^{p,q}(M,E) \qquad (1.11)$$

$H^{p,q}(M,E)$ is called the E-**valued Dolbeault cohomology group** of M of type (p,q), or simply the $\bar{\partial}$-**cohomology** of E of type (p,q).

E will not be referred to if $E \cong M \times \mathbb{C}$, i.e. if E is isomorphic to the trivial line bundle. Accordingly $H^{p,q}(M,E)$ will be denoted by $H^{p,q}(M)$ in such a case. Since we have assumed that any connected component of M admits a countable open basis, for any open covering $\mathscr{U}$ of M one can find a C^∞ partition of unity on M, say $\{\rho_\alpha\}$, such that $\{\operatorname{supp}\rho_\alpha\}$ is a refinement of $\mathscr{U}$. Therefore the sequence (1.9) is a **fine resolution**, i.e. a resolution by fine sheaves, of $\Omega^p(E)$. Since

$$H^k(U, \mathscr{C}^{p,q}(E)) = 0, \quad k \geq 1, \ p,q \geq 0$$

for any open set $U \subset M$, by the corollary of Theorem 1.18, similarly to Theorem 1.18 one has:

Theorem 1.31 (Dolbeault's isomorphism theorem).

$$H^{p,q}(M,E) \cong H^q(M, \Omega^p(E)).$$

For the detail of the proof, the reader is referred to [G-R] or [W].

Example 1.8. If M is a Stein manifold, Theorems 1.31 and 1.24 imply that

$$H^{p,q}(M,E) = 0, \quad q \geq 1$$

holds for any E.

1.4.2 Cohomology with Compact Support

Let X be a topological space. By definition, a **family of supports** on X is a collection of closed subsets of X, say Φ, satisfying the following two requirements.

(1) "$A \in \Phi$ and $K \subset A$" implies $\overline{K} \in \Phi$.
(2) $A, B \in \Phi \Rightarrow A \cup B \in \Phi$.

Let $\mathscr{F} \to X$ be a sheaf. We put

$$\mathscr{F}_\Phi(U) = \{s \in \mathscr{F}(U); \text{supp}\, s \in \Phi\}.$$

Since any homomorphism $\alpha : \mathscr{F} \to \mathscr{G}$ satisfies

$$\alpha_U(\mathscr{F}_\Phi(U)) \subset \mathscr{G}_\Phi(U),$$

one has a complex $\mathscr{F}_\Phi^*(X) := \{\mathscr{F}_\Phi^k(X), j_X^k\}_{k \geq 0}$ associated to the canonical flabby resolution of $\mathscr{F}$.

Definition 1.23. The p-th $\mathscr{F}$-valued **cohomology group of X supported in** Φ, denoted by $H_\Phi^p(X, \mathscr{F})$, is defined as the p-th cohomology group of the complex $\mathscr{F}_\Phi^*(X)$.

Let

$$\Phi_0 = \{K \subset X; K \text{ is compact}\}.$$

Then Φ_0 is obviously a family of supports. $\mathscr{F}_{\Phi_0}(X)$(resp. $H_{\Phi_0}^p(X, \mathscr{F})$) will be simply denoted by $\mathscr{F}_0(X)$(resp. $H_0^p(X, \mathscr{F})$). The following exact sequence is useful:

$$0 \to H_0^0(X, \mathscr{F}) \to H^0(X, \mathscr{F}) \to \text{ind.lim}_{K \Subset X} H^0(X \setminus K, \mathscr{F}) \to$$

$$\to H_0^1(X, \mathscr{F}) \to H^1(X, \mathscr{F}) \to \text{ind.lim}_{K \Subset X} H^1(X \setminus K, \mathscr{F}) \to \cdots.$$

Here $K \Subset X$ means that K is relatively compact in X. We put

$$C_0^{p,q}(M, E) = \{u \in C^{p,q}(M, E); \text{supp}\, u \Subset M\}$$

and

$$H_0^{p,q}(M, E) = \frac{\text{Ker}(\bar{\partial} : C_0^{p,q}(M, E) \to C_0^{p,q+1}(M, E))}{\text{Im}(\bar{\partial} : C_0^{p,q-1}(M, E) \to C_0^{p,q}(M, E))}.$$

Then, similarly to Theorem 1.31, given any holomorphic vector bundle $E \to M$ one has:

Theorem 1.32. $H_0^{p,q}(M, E) \cong H_0^q(M, \Omega^p(E))$.

We note that, combining the vanishing of $H^{p,q}(\mathbb{C}^n)$ for $q \geq 1$ with Theorem 1.1, one has $H_0^{p,1}(\mathbb{C}^n) = 0$ if $n \geq 2$. In fact, for any C^∞ $\bar{\partial}$-closed $(p,1)$-form v on $\mathbb{C}^n$ with compact support, there exists a C^∞ $(p,0)$-form u satisfying $\bar{\partial} u = v$ because $H^{p,1}(\mathbb{C}^n) = 0$, but there exists $f \in \mathscr{O}(\mathbb{C}^n)$ such that $f = u$ holds outside a compact subset of $\mathbb{C}^n$ by Theorem 1.1. Therefore v is the $\bar{\partial}$-image of a compactly supported function $u - f$. This argument can be generalized immediately to show that $H_0^{p,1}(D) = 0$ for any domain $D \subset \mathbb{C}^n$ $(n \geq 2)$ with unbounded and connected

complement. That $H_0^{p,q}(\mathbb{C}^n) = 0$ for $q \leq n-1$ can be shown similarly, but much
more general and straightforward reasoning is given by Serre's duality theorem
explained below.

1.4.3 Serre's Duality Theorem

The duality between the space of compactly supported C^∞ functions and the space
of distributions is carried over to the spaces of $\bar{\partial}$-cohomology groups. Such a duality
theorem holds on complex manifolds and can be extended on complex spaces after
an appropriate modification. We shall restrict ourselves to the duality on complex
manifolds here. For the duality theorem, an object of basic importance is the space
of currents. By definition, a **current of type** (p,q) on M, a (p,q)-**current** for short,
is an element of the (topological) dual space of $C_0^{n-p,n-q}(M)$, say $\mathscr{K}^{p,q}(M)$, where
the topology of $C_0^{n-p,n-q}(M)$ is that of the uniform convergence of all derivatives
with uniformly bounded supports. The topology of $\mathscr{K}^{p,q}(M)$ is defined as that of
the uniform convergence on bounded sets (the strong dual topology). Similarly, the
space $\mathscr{K}_0^{p,q}(M)$ of compactly supported (p,q)-currents is defined as

$$\mathscr{K}_0^{p,q}(M) = \{u \in \mathscr{K}^{p,q}(M); \operatorname{supp} u \Subset M\}.$$

For any holomorphic vector bundle E over M, the space $\mathscr{K}^{p,q}(M,E)$ of E-valued
(p,q)-currents is similarly defined as the dual of the space of $C_0^{n-p,n-q}(M,E^*)$.
$\mathscr{K}_0^{p,q}(M,E)$ is defined as well. $C^{p,q}(M,E)$ is naturally identified with a subset of
$\mathscr{K}^{p,q}(M,E)$. Since the Dolbeault complex with Dolbeault's lemma is naturally
extended to the complex of sheaves of the germs of currents, which are obviously
fine, one has canonical isomorphisms

$$H^{p,q}(M,E) \cong \frac{\operatorname{Ker}(\bar{\partial} : \mathscr{K}^{p,q}(M,E) \to \mathscr{K}^{p,q+1}(M,E))}{\operatorname{Im}(\bar{\partial} : \mathscr{K}^{p,q-1}(M,E) \to \mathscr{K}^{p,q}(M,E))}$$

and

$$H_0^{p,q}(M,E) \cong \frac{\operatorname{Ker}(\bar{\partial} : \mathscr{K}_0^{p,q}(M,E) \to \mathscr{K}_0^{p,q+1}(M,E))}{\operatorname{Im}(\bar{\partial} : \mathscr{K}_0^{p,q-1}(M,E) \to \mathscr{K}_0^{p,q}(M,E))}.$$

The pairing

$$\mathscr{K}^{p,q}(M,E) \times C_0^{n-p,n-q}(M,E^*) \to \mathbb{C}$$

is compatible with the exterior derivatives so that from the complexes

$$\mathscr{K}^{p,\cdot}(M,E) : 0 \to \mathscr{K}^{p,0}(M,E) \to \mathscr{K}^{p,1}(M,E) \to \cdots$$

and

$$C_0^{n-p,\cdot}(M,E^*) : 0 \to C_0^{n-p,0}(M,E^*) \to C_0^{p,1}(M,E^*) \to \cdots$$

a pairing

$$H^{p,q}(M,E) \times H_0^{n-p,n-q}(M,E^*) \to \mathbb{C}$$

is induced. Therefore one has a canonically defined continuous linear map

$$\iota^{p,q} : H_0^{n-p,n-q}(M,E^*) \to (H^{p,q}(M,E))',$$

where V' denotes for any locally convex space V the dual equipped with the strong topology. The map $\iota^{p,q}$ is surjective. To see this, first observe that any $\eta \in (H^{p,q}(M,E))'$ lifts to a continuous linear map from $C^{p,q}(M,E) \cap \mathrm{Ker}\bar{\partial}$ to $\mathbb{C}$, so that it also lifts to an element $\tilde{\eta}$ of $(C^{p,q}(M,E))' = \mathscr{K}_0^{n-p,n-q}(M,E^*)$. Since $\tilde{\eta}$ vanishes on the image of $\bar{\partial} : C^{p,q-1}(M,E) \to C^{p,q}(M,E)$, one has $\bar{\partial}\tilde{\eta} = 0$. Hence $\iota^{p,q}(\tilde{\eta}) = \eta$.

Similarly, we have natural surjective linear maps

$$\iota_0^{p,q} : H^{p,q}(M,E) \to (H_0^{n-p,n-q}(M,E^*))'$$

induced by the pairing

$$\mathscr{K}_0^{n-p,n-q}(M,E) \times C^{p,q}(M,E^*) \to \mathbb{C}.$$

Serre's duality theorem describes a necessary and sufficient condition for the maps $\iota^{p,q}$ and $\iota_0^{p,q}$ to be topological isomorphisms. Here the dual spaces $(H^{p,q}(M,E))'$ and $(H_0^{n-p,n-q}(M,E^*))'$ are equipped with the topology of uniform convergence on bounded sets.

Theorem 1.33. *The following are equivalent:*

(1) $\iota^{p,q}$ *is a topological isomorphism.*
(2) $\iota_0^{p,q+1}$ *is a topological isomorphism.*
(3) $H^{p,q+1}(M,E)$ *is a Hausdorff space.*
(4) $H_0^{n-p,n-q}(M,E^*)$ *is a Hausdorff space.*
(5) $\mathrm{Im}(\bar{\partial} : \mathscr{K}^{p,q}(M,E) \to \mathscr{K}^{p,q+1}(M,E)) =$
    $\{f \in \mathscr{K}^{p,q+1}(M,E); \langle f,g \rangle = 0 \text{ for any } g \in C_0^{n-p,n-q-1}(M,E^*) \cap \mathrm{Ker}\bar{\partial}\}.$
(6) $\mathrm{Im}(\bar{\partial} : C_0^{n-p,n-q-1}(M,E^*) \to C_0^{n-p,n-q}(M,E^*)) =$
    $\{g \in C_0^{n-p,n-q}(M,E^*); \langle f,g \rangle = 0 \text{ for any } f \in \mathscr{K}_0^{p,q}(M,E) \cap \mathrm{Ker}\bar{\partial}\}.$

Proof. It is standard that (5) and (6) are equivalent. Indeed, given two reflexive locally convex vector spaces say A and B, a continuous linear map $\alpha : A \to B$ and its transpose $\alpha'; B' \to A'$, we have an equivalence

$$\mathrm{Im}\,\alpha = \overline{\mathrm{Im}\,\alpha} \iff \mathrm{Im}\,\alpha' = \overline{\mathrm{Im}\,\alpha'}.$$

Equivalences (3) $\iff$ (5) and (4) $\iff$ (6) are obvious.

(5) $\Rightarrow$ (1): By Banach's open mapping theorem, it suffices to show that $\iota^{p,q}$ is bijective. Since the proof of surjectivity is over, it remains to show the injectivity. Suppose that $\iota^{p,q}([v]) = 0$ for some $v \in C^{n-p,n-q}(M, E^*) \cap \mathrm{Ker}\bar{\partial}$. Then $\langle u, v \rangle = 0$ for any $u \in \mathscr{H}^{p,q}(M, E) \cap \mathrm{Ker}\bar{\partial}$. Since $\mathrm{Im}(\bar{\partial} : \mathscr{H}^{p,q}(M, E) \to \mathscr{H}^{p,q+1}(M, E))$ is closed by (5), by Banach's open mapping theorem one can find a continuous linear map

$$w : \mathrm{Im}(\bar{\partial} : \mathscr{H}^{p,q}(M, E) \to \mathscr{H}^{p,q+1}(M, E)) \longrightarrow \mathbb{C}$$

such that

$$w(\bar{\partial}u) = \langle u, v \rangle \ \text{for any } u \in \mathscr{H}^{p,q}(M, E).$$

Therefore, by the Hahn–Banach theorem there exists a $\tilde{w} \in \mathscr{H}^{p,q+1}(M, E)' = C_0^{n-p,n-q-1}(M, E^*)$ such that $\langle \bar{\partial}u, \tilde{w} \rangle = \langle u, v \rangle$ holds for all $u \in \mathscr{H}^{p,q}(M, E)$, which means $(-1)^{\deg u+1}\bar{\partial}\tilde{w} = v$, so that $[v] = 0$.

(6) $\Rightarrow$ (2): Similar to the above.

(1) $\Rightarrow$ (4): Since $H^{p,q}(M, E)'$ is Hausdorff, (1) implies that $H_0^{n-p,n-q}(M, E^*)$ is Hausdorff. The proof of (2) $\Rightarrow$ (5) is similar.

Thus we have shown

$$(5) \iff (3)$$

and

$$(5) \Rightarrow (1) \Rightarrow (4) \iff (6) \Rightarrow (2) \Rightarrow (3) \iff (5).$$

Hence (1) $\sim$ (6) are all equivalent.   □

By the unique continuation theorem for analytic functions, obviously $H_0^{p,0}(M, E) = 0$ holds for any $p \geq 0$ if M is connected and noncompact. Hence Serre's duality theorem implies that $(H^{p,n}(M, E))' = 0$ ($p \geq 0$) holds for any connected noncompact complex manifold M. But actually, $H^{p,n}(M, E) = 0$ in such a situation (cf. [Siu]). Exploiting this fact and the Serre duality, let us note some examples of non-Hausdorff cohomology.

Example 1.9. If $M = \mathbb{C}^2 \setminus (\mathbb{R} \times \{0\})$, $H^{0,1}(M)$ and $H_0^{0,2}(M)$ are not Hausdorff. In fact, if $H^{0,1}(M)$ were Hausdorff, since $H^{0,2}(M) = 0$ as above, Serre's duality theorem and the remark after Theorem 1.32 would imply that $H^{0,1}(M) = 0$. But any domain $D \subset \mathbb{C}^2$ with $H^{0,1}(D) = 0$ is a domain of holomorphy, because every holomorphic function on $D \cap \{z_1 = const\}$ can be holomorphically extended to D in this situation. But $\mathbb{C}^2 \setminus (\mathbb{R} \times \{0\})$ is not a domain of holomorphy, as is easily seen from Theorem 1.1. Therefore $H^{0,1}(M) \neq 0$, which is a contradiction. Further, since $H^{0,1}(M)$ is not Hausdorff, it follows that $H_0^{2,2}(M) \ncong (H^{0,0}(M))'$.

1.4.4 Fiber Metric and L^2 Spaces

Let M be a complex manifold of dimension n and let $E \to M$ be a holomorphic vector bundle of constant rank r. By a **fiber metric** of E we shall mean a collection of positive definite Hermitian forms on the fibers E_x ($x \in M$) which is of class C^∞ as a section of $\mathrm{Hom}(E, \overline{E^*})$. A **Hermitian metric** on M is by definition a fiber metric of the holomorphic tangent bundle $T_M^{1,0}$. Since M has a countable basis of open sets, fiber metrics of E can be constructed by patching locally defined fiber metrics of $E|_U$ ($U \subset M$) by a C^∞ partition of unity. For any fiber metric $h \in C^\infty(M, \mathrm{Hom}(E, \overline{E^*}))$, a *twist*

$$\partial_h : C^{p,q}(M, E) \longrightarrow C^{p+1,q}(M, E)$$

of $\partial : C^{p,q}(M) \to C^{p+1,q}(M)$ is defined by

$$\partial_h = h^{-1} \circ \partial \circ h.$$

The operator $D_h = \partial_h + \bar{\partial}$ is called the **Chern connection** of (E, h). It is easy to see that $D_h{}^2$ is naturally identified with the exterior multiplication by a $\mathrm{Hom}(E, E)$-valued $(1, 1)$ form, say Θ_h from the left–hand side. The cohomology class represented by $\frac{1}{2\pi i}\Theta_{\det h} = \frac{1}{2\pi i}\mathrm{Tr}\Theta_h$ in $H^2(M, \mathbb{Z})$ is called the **first Chern class** of E and denoted by $c_1(E)$. If, moreover, M is compact and $\dim M = 1$, we put

$$\deg E = \frac{1}{2\pi i} \int_M \mathrm{Tr}\Theta_h.$$

and call it the **degree** of E. The degree is a topological invariant of E. This notion is generalized to the bundles over complex curves and further to higher–dimensional cases by fixing a set of divisors. However, we shall not go into this aspect of the theory of vector bundles in subsequent chapters. (See [Kb-2] for these materials.)

For any trivialization

$$E|_U \ni \xi \mapsto (\pi(\xi), \psi(\xi)) \in U \times \mathbb{C}^r$$

with $\psi(\xi) = (\xi^1, \ldots, \xi^r)$, the length $|\xi|_h$ of ξ with respect to h is expressed as

$$|\xi|_h^2 = \psi(\xi) h_U{}^t\overline{\psi(\xi)}$$

for some matrix valued C^∞ function h_U on U. Hence a fiber metric of E is naturally identified with a system of matrix–valued C^∞ functions h_j on U_j ($M = \bigcup U_j$ and $E_{U_j} \cong U_j \times \mathbb{C}^r$) such that $h_j(x)$ are positive definite Hermitian matrices and $h_j = {}^te_{kj}h_k\overline{e}_{kj}$ is satisfied on $U_j \cap U_k$ for the system of transition functions e_{jk} associated to the local trivializations of $E|_{U_j}$. A holomorphic vector bundle equipped with a fiber metrics is called a **Hermitian holomorphic vector bundle**. For a Hermitian holomorphic vector bundle (E, h), a local frame $s = (s_1, \ldots, s_r)$ of E defined on a neighborhood U of $x \in M$ is said to be **normal** at x if the matrix representation h_s of the fiber metric $h \in C^\infty(M, \mathrm{Hom}(E, \bar{E}^*))$ with respect to the local trivialization

$$E|_U \ni \sum_{j=1}^{r} \xi^j s_j(y) \mapsto (y, \xi_1, \ldots, \xi_r) \in U \times \mathbb{C}^r$$

satisfies

$$h_s(x) = \begin{pmatrix} 1 & 0 & \cdots & 0 \\ 0 & 1 & \cdots & 0 \\ \vdots & \vdots & \ddots & \vdots \\ 0 & \cdots & 0 & 1 \end{pmatrix} \quad \text{and} \quad dh_s(x) = 0.$$

It is easy to see that normal local frames exist for any (E, h) and $x \in M$. They are useful to check the validity of local formulas on the differential geometric quantities. Anyway, once we have a Hermitian metric on M and a fiber metric on E, the vector space $C_0^{p,q}(M, E)$ is naturally equipped with a topology of pre-Hilbert space which is much closer to the space we live in than those used in the proof of Serre's duality theorem. The purpose of the remaining four chapters is to make use of this advantage as far as possible.

Chapter 2
Analyzing the L^2 $\bar{\partial}$-Cohomology

Abstract For the bundle-valued differential forms on complex manifolds, a method of solving $\bar{\partial}$-equations with a control of L^2 norm is discussed. Basic results are existence theorems for such solutions under curvature conditions. They are variants of Kodaira's cohomology vanishing theorem on compact Kähler manifolds, and formulated as vanishing theorems with L^2 conditions. Some of these L^2 vanishing theorems are generalized to finite-dimensionality theorems under the assumptions on the bundle-convexity. Besides applications to holomorphic functions, extensions of the Hodge theory to noncompact manifolds will also be discussed.

2.1 Orthogonal Decompositions in Hilbert Spaces

The method of orthogonal projection introduced by H. Weyl [Wy-1] was an innovation in potential theory in the sense that it provided a general method of solving the Laplace equations without appealing to the fundamental solutions. This method has developed into a basic existence theory which is useful in complex analysis on complex manifolds. Its basic part can be stated in an abstract form that certain inequality implies the solvability of an equation with an estimate for the norms.

2.1.1 Basics on Closed Operators

Let $H_j(j = 1, 2)$ be two Hilbert spaces. Unless stated otherwise, we shall only consider complex Hilbert spaces. We shall denote by $(,)_j$ and $\| \|_j$ respectively the inner products and the norms of H_j. Later we shall use also the notations $(,)_{H_j}$ and $\| \|_{H_j}$. By a closed operator from H_1 to H_2, we mean a $\mathbb{C}$-linear map T from a dense linear subspace $\Omega \subset H_1$ to H_2 whose graph $G_T = \{(u, Tu) \in H_1 \times H_2 ; u \in \Omega\}$ is closed in $H_1 \times H_2$. Ω is called the **domain** of T and denoted by Dom$T$. The image $T(\Omega)$ of T will be denoted by ImT unless it is confused with the "imaginary part" of T. Accordingly, $\overline{\text{Im}T}$ stands for the closure of the image of T, and not for the conjugate of the imaginary part of T. The kernel $\{u; Tu = 0\}$ of T will be denoted by Ker$T$. Note that Ker$T$ is closed since so is G_T.

© Springer Japan 2015   41
T. Ohsawa, L^2 *Approaches in Several Complex Variables*, Springer Monographs in Mathematics, DOI 10.1007/978-4-431-55747-0_2

The adjoint of a closed operator T, denoted by T^*, is by definition a closed operator from H_2 to H_1 satisfying

$$G_{T^*} = \{(v, w) \in H_2 \times H_1; (v, Tu)_2 = (w, u)_1 \text{ for all } u \in \text{Dom}T\}. \tag{2.1}$$

Note that $\overline{G_{T^*}} = G_{T^*}$ because the right–hand side of (2.1) is $(G_{-T})^{\perp}$, the orthogonal complement of G_{-T} in $H_1 \times H_2$, up to the exchange of components. That $(v, w), (v, w') \in G_{T^*}$ implies $w = w'$ follows from $\overline{\text{Dom}T} = H_1$. Vice versa, that $\overline{\text{Dom}T^*} = H_2$ is because T is single-valued. Obviously $T^{**} = T$.

Proposition 2.1. $\text{Im}T^{\perp} = \text{Ker}T^*$.

Corollary 2.1. $H_2 = \overline{\text{Im}T} \oplus \text{Ker}T^*$.

Similarly, $H_1 = \overline{\text{Im}T^*} \oplus \text{Ker}T$, for $T^{**} = T$.

2.1.2   Kodaira's Decomposition Theorem and Hörmander's Lemma

Let $H_j (j = 1, 2, 3)$ be three Hilbert spaces with norms $\| * \|_j$. Let T be a closed operator from H_1 to H_2, and let S be a closed operator from H_2 to H_3 satisfying $\text{Dom}S \supset \text{Im}T$ and $ST = 0$. Then, by Proposition 2.1 one has

$$H_2 = \overline{\text{Im}T} \oplus \text{Ker}T^* = \overline{\text{Im}S^*} \oplus \text{Ker}S. \tag{2.2}$$

Since $ST = 0$, $\text{Im}T \subset \text{Ker}S$ so that $\overline{\text{Im}T} \subset \text{Ker}S$. Similarly $\overline{\text{Im}S^*} \subset \text{Ker}T^*$, since $T^*S^* = 0$ follows immediately from $ST = 0$. Hence $\overline{\text{Im}T}$ and $\overline{\text{Im}S^*}$ are orthogonal to each other. Combining these one has the following decomposition theorem first due to K. Kodaira [K-1, Theorem 5].

Theorem 2.1. $H_2 = \overline{\text{Im}T} \oplus \overline{\text{Im}S^*} \oplus (\text{Ker}S \cap \text{Ker}T^*)$.

In order to analyze this decomposition more in detail, the following is of basic importance.

Lemma 2.1 (cf. [Hö-2, Proof of Lemma 4.1.1]). *Let $v \in H_2$. Then $v \in \text{Im}T$ if and only if there exists a nonnegative number C such that*

$$|(u, v)_2| \leq C\|T^*u\|_1 \tag{2.3}$$

holds for any $u \in \text{Dom}T^*$. Moreover, the infimum of such C is $\min\{\|w\|_1; Tw = v\}$.

Proof. If $v = Tw$ for some $w \in H_1$, $|(u, v)_2| = |(u, Tw)_2| = |(T^*u, w)_1| \leq \|T^*u\|_1 \cdot \|w\|_1$. Hence one may put $C = \|w\|_1$. Conversely, suppose that (2.3) holds for any $u \in \text{Dom}T^*$. Then the correspondence $u \rightarrow (u, v)_2$ induces a continuous $\mathbb{C}$ linear map from $\text{Im}T^*$ to $\mathbb{C}$, and further, one from H_1 by composing the orthogonal projection $H_1 \to \mathrm{Im}T^*$. Therefore, there exists a $w \in H_1$ such that $\|w\|_1 \le C$ and $(u, v)_2 = (T^*u, w)_1$ holds for any $u \in \mathrm{Dom}T^*$. $\square$

Corollary 2.2. $\mathrm{Ker}S = \mathrm{Im}T$ *if and only if there exists a function* $C : \mathrm{Ker}S \to [0, \infty)$ *such that* $|(u, v)_2| \le C(v)\|T^*u\|_1$ *holds for any* $u \in \mathrm{Dom}T^*$ *and* $v \in \mathrm{Ker}S$.

Corollary 2.3. $\mathrm{Ker}S = \mathrm{Im}T$ *if there exists a constant* $C > 0$ *such that*

$$\|u\|_2 \le C(\|T^*u\|_1 + \|Su\|_3) \tag{2.4}$$

holds for any $u \in \mathrm{Dom}T^* \cap \mathrm{Dom}S$.

Proof. Suppose that (2.4) holds for any $u \in \mathrm{Dom}T^* \cap \mathrm{Dom}S$ and take any $v \in \mathrm{Ker}S$. Let $u \in \mathrm{Dom}T^*$ and let $u = u_1 + u_2, u_1 \in \mathrm{Ker}S, u_2 \in \mathrm{Im}S^*$ be the orthogonal decomposition. Then $T^*u = T^*u_1$, and $(u, v)_2 = (u_1, v)_2$ since $u_2 \perp v$. Hence $|(u, v)_2| = |(u_1, v)_2| \le C\|T^*u_1\| \cdot \|v\| = C\|T^*u\| \cdot \|v\|$, so that $v \in \mathrm{Im}T$ by Corollary 2.1. $\square$

Extension of Corollary 2.3 to the following is immediate.

Theorem 2.2. $H_2 = \mathrm{Im}T \oplus \mathrm{Im}S^* \oplus (\mathrm{Ker}S \cap \mathrm{Ker}T^*)$ *if there exists a constant* $C > 0$ *such that (2.4) holds for any* $u \in \mathrm{Dom}T^* \cap \mathrm{Dom}S \cap (\mathrm{Ker}S \cap \mathrm{Ker}T^*)^\perp$.

The hypothesis of Theorem 2.2 is fulfilled in a situation naturally arising in certain existence questions in 1.5 above. To see this, the following will be applied later.

Proposition 2.2. *Assume that from every sequence* $u_k \in \mathrm{Dom}T^* \cap \mathrm{Dom}S$ *with* $\|u_k\|_2$ *bounded and* $T^*u_k \to 0$ *in* H_1, $Su_k \to 0$ *in* H_3, *one can select a strongly convergent subsequence. Then (2.4) holds for some* $C > 0$ *and any* $u \in \mathrm{Dom}T^* \cap \mathrm{Dom}S \cap (\mathrm{Ker}S \cap \mathrm{Ker}T^*)^\perp$, *and* $\mathrm{Ker}S \cap \mathrm{Ker}T^*$ *is finite dimensional.*

The following is also applied later:

Theorem 2.3 (cf. Theorem 1.1.4 in [Hö-1]). *Let* F *be a closed subspace of* H_2 *containing* $\mathrm{Im}T$. *Assume that* $\|u\|_2 \le C(\|T^*u\|_1 + \|Su\|_3)$ *holds for any* $u \in \mathrm{Dom}T^* \cap \mathrm{Dom}S \cap F$. *Then:*

(i) *For any* $v \in \mathrm{Ker}S \cap F$ *one can find* $w \in \mathrm{Dom}T$ *such that* $Tw = v$ *and* $\|w\|_1 \le C\|v\|_2$.

(ii) *For any* $w \in \mathrm{Im}T^*$ *one can find* $v \in \mathrm{Dom}T^*$ *such that* $T^*v = w$ *and* $\|v\|_2 \le C\|w\|_1$.

Proof. (i) Let $v \in \mathrm{Ker}S \cap F$, let $u \in \mathrm{Dom}T^*$, and let $u = u_1 + u_2 + u_3$, where $u_1 \in \mathrm{Ker}S \cap F$, $u_2 \in \mathrm{Ker}S$ and $u_2 \perp F$, and $u_3 \perp \mathrm{Ker}S$. Since $\mathrm{Im}T \subset F$ and $u_2 \perp F$, $u_2 \perp \mathrm{Im}T$ so that $u_2 \in \mathrm{Ker}T^*$. Moreover, $u_3 \in \overline{\mathrm{Im}S^*} \subset \mathrm{Ker}T^*$. Therefore $u_2 + u_3 \in \mathrm{Ker}T^*$, so that $u_1 \in \mathrm{Dom}T^*$, and

$$|(u, v)_2| = |(u_1, v)_2| \le C\|T^*u_1\|_1 \cdot \|v\|_2 \tag{2.5}$$

Hence we have $|(u, v)_2| \leq C\|T^*u\|_1 \cdot \|v\|_2$ from (2.5). Therefore the linear functional $u \mapsto (u, v)_2$ on $\mathrm{Dom}T^*$ is continuous in T^*u, so that there exists $w \in H_1$ such that $\|w\|_1 \leq C\|v\|_2$ and

$$(u, v)_2 = (T^*u, w)_1$$

holds for every $u \in \mathrm{Dom}T^*$ so that $v = Tw$.

(ii) Let $w = T^*v_0$ and $v_0 = v_1 + v_2$, where $v_1 \perp \mathrm{Ker}T^*$ and $v_2 \in \mathrm{Ker}T^*$. Then $v_1 \in \overline{\mathrm{Im}T}$ so that $v_1 \in F$. Hence $v_1 \in F \cap \mathrm{Dom}T^* \cap \mathrm{Ker}S$ so that

$$\|v_1\|_2 \leq C\|T^*v_1\| = C\|T^*v_0\| = C\|w\|.$$

Thus it suffices to put $v = v_1$.   □

2.1.3 Remarks on the Closedness

Let the situation be as above. A basic observation of meta-theoretical importance is that $\mathrm{Im}T$ is closed if and only if $T|_{(\mathrm{Ker}T)^\perp}$ is invertible. In other words the following holds.

Proposition 2.3. *The following are equivalent:*

 (i) $\overline{\mathrm{Im}T} = \mathrm{Im}T$.
(ii) *There exists a constant C such that*
     $\|u\| \leq C\|Tu\|$ *holds for any* $u \in \mathrm{Dom}T \cap (\mathrm{Ker}T)^\perp$.

Proof. (ii) $\Rightarrow$ (i) is obvious. (i) $\Rightarrow$ (ii) follows from Banach's open mapping theorem, or closed graph theorem, or uniform boundedness theorem.   □

Combining Proposition 2.3 with Corollary 2.1 one has:

Theorem 2.4. *The following are equivalent:*

 (i) $\overline{\mathrm{Im}T} = \mathrm{Im}T$.
(ii) $\overline{\mathrm{Im}T^*} = \mathrm{Im}T^*$.

Accordingly, Corollary 2.2 is also strengthened to the following.

Theorem 2.5. $H_2 = \mathrm{Im}T \oplus \mathrm{Im}S^*$ *if and only if there exists a constant C such that (2.4) holds for any* $u \in \mathrm{Dom}T^* \cap \mathrm{Dom}S$.

Similarly, the converse of Theorem 2.2 also holds.

Let us add one more remark which is not so often mentioned but seems to be useful. For an application see Example 2.2 below. (See also [Oh-5].) The proof is left to the reader as an exercise.

Proposition 2.4. $\overline{\mathrm{Im}T} = \mathrm{Im}T$ *if* $\dim \mathrm{Ker}S/\mathrm{Im}T < \infty$.

2.2 Vanishing Theorems

Solvability criteria for $\bar{\partial}$-equations on complex manifolds are often described as cohomology vanishing theorems. In order to apply the abstract theory presented in the previous section, it is necessary to know that certain inequality holds for the bundle-valued differential forms under some curvature condition. The first vanishing theorem of this type was established by Kodaira [K-2] on compact Kähler manifolds and substantially by Oka [O-1, O-4] on pseudoconvex Riemann domains over $\mathbb{C}^n$. A vanishing theorem for L^2 $\bar{\partial}$-cohomology groups on complete Kähler manifolds unifies Kodaira's vanishing theorem and Cartan's Theorem B on Stein spaces. This viewpoint was first presented in a paper of Andreotti and Vesentini [A-V-1] and later effectively developed in [A-V-2]. Independently and more thoroughly, Hörmander [Hö-1] established the method of L^2 estimates for the $\bar{\partial}$-operator, extending also the preceding works of Morrey [Mry] and Kohn [Kn] on the complex boundary value problem. The advantage of this method is its flexibility in the limiting procedures as in Theorems 2.14 and 2.16. The argument below is based on the method of Andreotti and Vesentini. It will be refined in the next section to recover a finiteness theorem of Hörmander.

2.2.1   Metrics and L^2 $\bar{\partial}$-Cohomology

Let (M, ω) be a (not necessarily connected but pure dimensional) Hermitian manifold of dimension n and let (E, h) be a Hermitian holomorphic vector bundle over M. In order to analyze the $\bar{\partial}$-cohomology groups of (M, E), the metric structure (ω, h) is useful. As before, we denote by $C^{p,q}(M, E)$ the set of E-valued C^∞ (p, q)-forms on M and by $C_0^{p,q}(M, E)$ the subset of $C^{p,q}(M, E)$ consisting of compactly supported forms.

The pointwise length of $u \in C^{p,q}(M, E)$ with respect to the fiber metric induced by ω and h, measured by regarding u as a section of $\wedge^p(T_M^{1,0})^* \otimes \wedge^q(T_M^{0,1})^* \otimes E$, is denoted by $|u|(= |u|_{\omega,h})$. The pointwise inner product of u and v is denoted by $\langle u, v \rangle(= \langle u, v \rangle_{\omega,h})$. Then the L^2 norm of u denoted by $\|u\|_h$, or simply by $\|u\|$, is defined as the square root of the integral

$$\int_M |u|^2 \frac{\omega^n}{n!} \tag{2.6}$$

which is finite if $u \in C_0^{p,q}(M, E)$. The inner product of u and v associated to the norm is denoted by $(u, v)_h$. $(u, v)_h$ is

$$\int_M \langle u, v \rangle_{\omega,h} \frac{\omega^n}{n!}$$

or

$$\frac{1}{2}\left(\|u+v\|^2 - \|u\|^2 - \|v\|^2\right) - \frac{i}{2}\left(\|iu+v\|^2 - \|u\|^2 - \|v\|^2\right) \tag{2.7}$$

by definition, but has an expression more convenient for computation. Namely,

$$(u,v)_h = \int_M u \wedge \overline{h * v}. \tag{2.8}$$

Here h is identified with a section of $E^* \otimes \bar{E}^* (\cong \mathrm{Hom}(E, \bar{E}^*))$ and $*$ is a map from $C^{p,q}(M,E)$ to $C^{n-q,n-p}(M,E)$ induced from the unique isometric bundle morphism $*$ between $\wedge^r(T^{\mathbb{C}}M)^*$ and $\wedge^{2n-r}(T^{\mathbb{C}}M)^*$ $(r = p + q)$ that satisfies

$$e_1 \wedge e_2 \wedge \cdots \wedge e_r \wedge \overline{*(e_1 \wedge e_2 \wedge \cdots \wedge e_r)} = |e_1 \wedge e_2 \wedge \cdots \wedge e_r|^2 \omega^n/n! \tag{2.9}$$

for all $e_j (1 \leq j \leq r)$ in a fiber of $(T_M^{\mathbb{C}})^*$. The map $*$ is called **Hodge's star operator**. For simplicity we put $\overline{*v} = \bar{*}v$. Then $\bar{*}$ is a map from $C^{p,q}(M,E)$ to $C^{n-p,n-q}(M,E)$.

Example 2.1. For $M = \mathbb{C}^n$ and $\omega = \frac{i}{2}\sum_{j=1}^n dz_j \wedge d\bar{z}_j$,

$$\bar{*}(dz_I \wedge d\bar{z}_J) = c_{IJ} dz_{I'} \wedge d\bar{z}_{J'}, \tag{2.10}$$

where I' and J' complement I and J, respectively, and $c_{IJ} = (-1)^{(n-p)q} i^{n^2} 2^{p+q-n}$, where $p = |I|$ and $q = |J|$.

Let $L^{p,q}_{(2)}(M,E)$ be the completion of the pre-Hilbert space $C^{p,q}_0(M,E)$ with respect to the L^2 norm. By Lebesgue's theory of integration, $L^{p,q}_{(2)}(M,E)$ is naturally identified with a subset of E-valued (p,q)-forms with locally square integrable $(= L^2_{loc})$ coefficients. Then every element f of $L^{p,q}_{(2)}(M,E)$ is naturally identified with a $\mathbb{C}$-linear function on $C^{p,q}_0(M,E)$ by the inner product $\cdot \to (\cdot,f)_h$. For simplicity, $\bar{\partial}$ will also stand for a densely defined map from $L^{p,q}_{(2)}(M,E)$ to $L^{p,q+1}_{(2)}(M,E)$ whose domain of definition, denoted by Dom$\bar{\partial}$, is $\{f \in L^{p,q}_{(2)}(M,E); \bar{\partial}f \in L^{p,q+1}_{(2)}(M,E)\}$, where $\bar{\partial}f$ is defined in the sense of distribution for any $f \in L^{p,q}_{(2)}(M,E)$. In other words, $\bar{\partial}f$ is regarded as an element of $C^{n-p,n-q-1}_0(M,E^*)^*$ by the equality

$$\int_M \bar{\partial}f \wedge v \left(= \bar{\partial}f(v)\right) = (-1)^{p+q+1}\int_M f \wedge \bar{\partial}v$$

$$\text{for all } v \in C^{n-p,n-q-1}_0(M,E^*), \tag{2.11}$$

and "$\bar{\partial}f \in L^{p,q+1}_{(2)}(M,E)$" means that there exists a unique element $w \in L^{p,q+1}_{(2)}(M,E)$ such that $\int_M w \wedge v = (-1)^{p+q+1}\int_M f \wedge \bar{\partial}v$ holds for any $v \in C^{n-p,n-q-1}_0(M,E^*)$.

We define the L^2 $\bar{\partial}$-**cohomology groups** $H_{(2)}^{p,q}(M,E)$ by

$$H_{(2)}^{p,q}(M,E) := \operatorname{Ker}\bar{\partial} \cap L_{(2)}^{p,q}(M,E)/\operatorname{Im}\bar{\partial} \cap L_{(2)}^{p,q}(M,E). \tag{2.12}$$

$L_{(2)}^{p,q}(M,E)$ and $H_{(2)}^{p,q}(M,E)$ will be denoted by $L_{(2)}^{p,q}(M,E,\omega,h)$ and $H_{(2)}^{p,q}(M,E,\omega,h)$, respectively, whenever (ω,h) must be visible. L^2 de Rham cohomology groups $H_{(2)}^r(M)$ are defined similarly. $\bar{\partial}$ is obviously a closed operator from $L_{(2)}^{p,q}(M,E)$ to $L_{(2)}^{p,q+1}(M,E)$ so that it has its adjoint. It will be denoted by $\bar{\partial}_h^*$, or more simply by $\bar{\partial}^*$. A basic fact is that $H_{(2)}^{p,q}(M,E) \cong \operatorname{Ker}\bar{\partial} \cap \operatorname{Ker}\bar{\partial}^* \cap L_{(2)}^{p,q}(M,E)$ if $\operatorname{Im}\bar{\partial} \cap L_{(2)}^{p,q}(M,E)$ is closed (cf. Theorem 2.1).

Example 2.2. With respect to the Euclidean metric,

$$\dim H_{(2)}^{p,q}(\mathbb{C}^n) = \begin{cases} 0 & \text{if } q=0 \text{ or } q>n, \\ \infty & \text{otherwise}, \end{cases}$$

for any $n \in \mathbb{N}$.

Indeed, $H_{(2)}^{p,0}(\mathbb{C}^n) = 0$ follows from Cauchy's estimate. That $H_{(2)}^{p,q}(\mathbb{C}^n) = 0$ for $q > n$ is trivial. To see that $\dim H_{(2)}^{p,1}(\mathbb{C}^n) = \infty$, it suffices to apply Propositions 2.3 and 2.4, combining $H_{(2)}^{p,0}(\mathbb{C}^n) = 0$ with an obvious fact that one can find a sequence $u_k \in L_{(2)}^{p,0}(\mathbb{C}^n)$ such that $\|u_k\| = 1$ and $\|\bar{\partial}u_k\| \to 0$ as $k \to \infty$. The infinite dimensionality for general q follows similarly. Namely, the non-Hausdorff property of $H_{(2)}^{p,q}(\mathbb{C}^n)$ for $2 \le q \le n$ follows from that there exists a sequence $u_k \in C_0^{p,q}(\mathbb{C}^n)$ such that $\bar{\partial}^* u_k (\perp \operatorname{Ker}\bar{\partial})$ are of norm 1 but $\|\bar{\partial}\bar{\partial}^* u_k\| \to 0$ as $k \to \infty$, which is also obvious as in the case $q = 1$.

To obtain more advanced results, one needs to find natural conditions on (ω, h) in order to apply abstract existence theorems in Sect. 2.1.2. An effective condition acceptable in most cases is the completeness of ω which guarantees in particular the density of $C_0^{p,q}(M,E)$ in $\operatorname{Dom}\bar{\partial}^* \cap L_{(2)}^{p,q}(M,E)$ with respect to the graph norm of $\bar{\partial}^*$, which will be explained below.

2.2.2 Complete Metrics and Gaffney's Theorem

A Hermitian manifold (M,ω) is said to be **complete** if M is complete as a metric space with respect to the distance associated to ω. Recall that the distance between $x, y \in M$ with respect to ω is defined as the infimum of $\int_0^1 \sqrt{\gamma^* g}$ where g is the fiber metric of $T_M^{1,0}$ associated to ω regarded as a section of $(T_M^{1,0})^* \otimes (T_M^{0,1})^*$ and γ runs through C^∞ maps from [0,1] to M satisfying $\gamma(0) = x$ and $\gamma(1) = y$. This distance will be denoted by $\operatorname{dist}_\omega(x,y)$, or simply by $d(x,y)$.

Example 2.3. $(\mathbb{C}^n, \frac{i}{2} \sum dz_j \wedge d\overline{z_j})$ is complete.

Proposition 2.5. *(M, ω) is complete if and only if $\{y; d(x, y) < R\}$ is relatively compact for any $x \in M$ and $R > 0$.*

Proof. The "if" part is obvious. The converse is easy to see from the Bolzano–Weierstrass theorem. □

Since $d(x, y)$ is Lipschitz continuous on $M \times M$, it can be approximated uniformly by a C^∞ function say $\tilde{d}(x, y)$ with bounded gradient. Let us fix a point $x_0 \in M$ and put $\rho(x) = \tilde{d}(x_0, x)$. Let $\chi : \mathbb{R} \to [0, \infty)$ be a C^∞ function such that

(i) $\chi|(-\infty, 1) \equiv 1$
    and
(ii) $\mathrm{supp}\chi \subset (-\infty, 2]$.

Then we put $\chi_R(x) = \chi\left(\frac{\rho(x)}{R}\right)$ for $R > 1$. An important property of χ_R is that $|d\chi_R| \leq C/R$ holds for some $C > 0$.

Proposition 2.6. *Let (M, ω) be a Hermitian manifold and let (E, h) be a Hermitian holomorphic vector bundle over M. Then, for any $u \in \mathrm{Dom}\bar{\partial} \cap L_{(2)}^{p,q}(M, E)$, $\chi_R u \in \mathrm{Dom}\bar{\partial} \cap L_{(2)}^{p,q}(M, E)$ and $\|\chi_R u - u\| + \|\bar{\partial}(\chi_R u) - \bar{\partial}u\| \to 0$ as $R \to \infty$.*

Proof. That $\bar{\partial}(\chi_R u) = \bar{\partial}\chi_R \wedge u + \chi_R \bar{\partial}u \in L_{(2)}^{p,q+1}(M, E)$ and $\lim_{R \to \infty} \|\chi_R u - u\| = 0$ is obvious. In order to see that $\lim_{R \to \infty} \|\bar{\partial}(\chi_R u) - \bar{\partial}u\| = 0$, it suffices to combine $\lim_{R \to \infty} \|\chi_R \bar{\partial}u - \bar{\partial}u\| = 0$ and $\lim_{R \to \infty} (\sup_M |\bar{\partial}\chi_R|) = 0$. □

If (M, ω) is complete, $\mathrm{supp}(\chi_R u) \Subset M$ for all R. Hence there exists for each R a sequence $u_k \in C_0^{p,q}(M, E)$ satisfying $\|u_k - \chi_R u\| + \|\bar{\partial}u_k - \bar{\partial}(\chi_R u)\| \to 0$ as $k \to \infty$. Combining this observation with Proposition 2.6, we have:

Proposition 2.7. *Let (M, ω) be a complete Hermitian manifold and let (E, h) be a Hermitian holomorphic vector bundle over M. Then $C_0^{p,q}(M, E)$ is dense in $\mathrm{Dom}\bar{\partial}$ with respect to the norm $\|u\| + \|\bar{\partial}u\|$.*

Similarly, since $\bar{\partial}^*$ acts on $C_0^{p,q+1}(M, E)$ as a differential operator $-\bar{*}h^{-1}\bar{\partial}\bar{h}\bar{*}$, which is easy to see from the Stokes' formula and $**u = (-1)^{\deg u}u$, Proposition 2.7 can be extended to the following important result which is first due to Gaffney [Ga] for the exterior derivative d and formulated for $\bar{\partial}$ by Andreotti and Vesentini in [A-V-1, A-V-2].

Theorem 2.6. *In the situation of Proposition 2.7, $C_0^{p,q}(M, E)$ is dense in $\mathrm{Dom}\bar{\partial} \cap \mathrm{Dom}\bar{\partial}^*$ with respect to the norm $\|u\| + \|\bar{\partial}u\| + \|\bar{\partial}^*u\|$.*

The importance of Theorem 2.6 for our purpose lies in that integration by parts is available without worrying about the boundary terms to obtain the estimates implying the existence theorems. To proceed in this way, formulas in the $C^\infty(M)$ algebra of operators on $\oplus_{p+q=0}^{2n} C^{p,q}(M, E)$ are useful. They will be described next.

2.2.3 Some Commutator Relations

Before presenting formulas involving $\bar{\partial}$, let us prepare some abstract formalism. Let $\mathscr{R}$ be a commutative ring and let $\mathscr{M}$ be a graded $\mathscr{R}$ module, i.e. $\mathscr{M}$ is a direct sum of submodules say $\mathscr{M}_j (j \in \mathbb{Z})$. If $u \in \mathscr{M}_j - \{0\}$, j is called the **degree** of u and denoted by $\deg u$. Let

$$\Pi_k(\mathscr{M}) = \{T \in \mathscr{M}^{\mathscr{M}}; T(\mathscr{M}_j) \subset \mathscr{M}_{j+k} \text{ for all } j\}.$$

For any $T \in \Pi_k(\mathscr{M}) - \{0\}$ we put $\deg T = k$. Then $\oplus_{k \in \mathbb{Z}} \Pi_k(\mathscr{M})$ is a graded left $\mathscr{R}$ algebra whose product is defined by composition. Elements of $\cup_{k \in \mathbb{Z}} \Pi_k(\mathscr{M})$ are said to be homogeneous. Given $S \in \Pi_k(\mathscr{M})$ and $T \in \Pi_\ell(\mathscr{M})$, we define the graded commutator of S and T by

$$[S, T]_{gr} = S \circ T - (-1)^{\deg S \deg T} T \circ S, \tag{2.13}$$

where we put $\deg 0 = 0$. The following straightforward consequence of the definition is very important.

Lemma 2.2 (Jacobi's identity). *For any homogeneous $S, T, U \in \Pi(\mathscr{M})$,*

$$[[S, T]_{gr}, U]_{gr} - [S, [T, U]_{gr}]_{gr} = (-1)^{\deg S \deg T + 1}[T, [S, U]_{gr}]_{gr}. \tag{2.14}$$

Now let $(M, \omega)(\dim M \geq 1)$ and (E, h) be as before and set $\mathscr{R} = C^\infty(M)$, $\mathscr{M} = \oplus_{p+q=0}^{2n} C^{p,q}(M, E)$ and $\mathscr{M}_j = \oplus_{p+q=j} C^{p,q}(M, E)$. Then, with respect to this natural grading, $\bar{\partial} \in \Pi_1(\mathscr{M})$. We shall identify the elements of $C^{p,q}(M, E)$ with those in $\Pi_{p+q}(\mathscr{M})$ by letting them act on $\mathscr{M}$ by exterior multiplication from the left–hand side. Given $\theta \in C^{p,q}(M)$, we define $\theta^* \in \Pi_{-p-q}(\mathscr{M})$ by requiring the equality $\langle \theta \wedge u, v \rangle = \langle u, \theta^* v \rangle$ for the pointwise inner product $\langle \cdot, \cdot \rangle = \langle \cdot, \cdot \rangle_h$ to hold for any $u \in C^{r,s}(M, E)$ and $v \in C^{r+p,s+q}(M, E)$. Since $\omega \in C^{1,1}(M), \omega^* \in \Pi_{-2}(\mathscr{M})$. We shall use the notation Λ for ω^* following [W-1, W-2] and [N-1]. Some formulas involving Λ are of special importance. They will be recalled below.

For the special case $(M, \omega) = (\mathbb{C}^n, \frac{i}{2} \sum dz_j \wedge d\bar{z}_j)$, it is easy to see that

$$\Lambda = \frac{1}{2i} \sum (d\bar{z}_j)^* (dz_j)^*$$

and

$$[d\bar{z}_j, \Lambda]_{gr} = -i(dz_j)^* (1 \leq j \leq n).$$

Since this formula can be applied pointwise, one has the following in general.

Lemma 2.3. *For any $\theta \in C^{0,1}(M)$, $[\theta, \Lambda]_{gr} = -i\bar{\theta}^*$.*

Proposition 2.8. *For any $\sigma, \tau \in C^{0,1}(M)$, $[\sigma, \tau^*]_{gr} + [\bar{\sigma}^*, \bar{\tau}]_{gr} = 0$.*

Proof. Since $\tau^* = i[\bar{\tau}, \Lambda]_{gr}$ and $\bar{\sigma}^* = -i[\sigma, \Lambda]_{gr}$, one has $[\sigma, \tau^*]_{gr} + [\bar{\sigma}^*, \bar{\tau}]_{gr} = i[\sigma, [\bar{\tau}, \Lambda]_{gr}]_{gr} + i[[\sigma, \Lambda]_{gr}, \bar{\tau}]_{gr} = [i[\sigma, \bar{\tau}]_{gr}, \Lambda]_{gr} = 0.$ $\qquad\square$

Similarly, replacing θ by $\bar{\partial}$ one has a useful expression for $[\bar{\partial}, \Lambda]_{gr}$. To describe it, let us put

$$\partial^* = -\bar{*}\partial\bar{*}. \tag{2.15}$$

Proposition 2.9. *For any* $u \in C_0^{p,q}(M)$ *and* $v \in C^{p+1,q}(M)$, $\int_M \partial u \wedge \bar{*}v = \int_M u \wedge \bar{*}\partial^* v$.

Proof. $\int_M \partial u \wedge \bar{*}v = \int_M d(u \wedge \bar{*}v) - (-1)^{p+q} \int_M u \wedge d\bar{*}v = -(-1)^{\deg v - 1} \int_M u \wedge \partial\bar{*}v = -\int_M u \wedge \bar{*}\bar{*}\partial\bar{*}v = \int_M u \wedge \bar{*}(-\bar{*}\partial\bar{*}v).$ $\qquad\square$

Lemma 2.4. $[\bar{\partial}, \Lambda]_{gr} = i\partial^*$ *if* $d\omega = 0$.

Proof. Since $\bar{\partial}$ and ∂^* are differential operators of the first order, it suffices to show the assertion for the Euclidean case. In this situation, first note that for any $u \in C^\infty(\mathbb{C}^n)$,

$$[\bar{\partial}, \Lambda]_{gr}(u dz_I \wedge d\overline{z_J}) = \frac{1}{i} \sum_{j \in I} \frac{\partial u}{\partial \overline{z_j}} dz_j^*(dz_I \wedge d\overline{z_J}). \tag{2.16}$$

Hence, if $v \in C_0^\infty(\mathbb{C}^n)$ and $\{j, K\} = I$,

$$([\bar{\partial}, \Lambda]_{gr}(u dz_I \wedge d\overline{z_J}), v) = \frac{1}{i} \int_{\mathbb{C}^n} \sum_{j=1}^n \frac{\partial u}{\partial \overline{z_j}} dz_j^*(dz_I \wedge d\overline{z_J}) \wedge \bar{*}(v dz_K \wedge d\overline{z_J})$$

$$= \frac{1}{i} \int_{\mathbb{C}^n} \frac{\partial u}{\partial \overline{z_j}} v \cdot i^{n^2} (dz_1 \wedge \cdots \wedge dz_n \wedge d\overline{z_1} \wedge \cdots \wedge d\overline{z_n})$$

$$= -\frac{1}{i} \int_{\mathbb{C}^n} u \frac{\partial v}{\partial \overline{z_j}} i^{n^2} (dz_1 \wedge \cdots \wedge dz_n \wedge d\overline{z_1} \wedge \cdots \wedge d\overline{z_n})$$

$$= i \int_{\mathbb{C}^n} u dz_I \wedge d\overline{z_J} \wedge \bar{*}(\sum \frac{\partial v}{\partial \overline{z_j}} dz_j \wedge dz_K \wedge d\overline{z_J}) = i(u, \partial v)(= i(u, \partial v)_\omega).$$

Hence $[\bar{\partial}, \Lambda]_{gr} = i\partial^*$. $\qquad\square$

For the sake of consistency, we shall denote $\bar{\partial}^*$ (resp. $\bar{\partial}_h^*$) by $\bar{\partial}^*$ (resp. $\bar{\partial}_h^*$) when it operates on $C^{p,q}(M)$ (resp. $C^{p,q}(M, E)$) as a differential operator. Let $\partial_h = h^{-1} \circ \partial \circ h$. Then $[\bar{\partial}, \Lambda]_{gr} + i\partial^*$ and $[\partial_h, \Lambda]_{gr} - i\bar{\partial}_h^*$ are operators of order zero. (For the explicit expressions of them, see [Dm-3] or [Oh-7].)

Proposition 2.10. $[\bar{\partial}, \bar{\partial}^*]_{gr} - [\partial^*, \partial]_{gr} = 0$ *if* $d\omega = 0$.

Proof. Since $\partial^* = -i[\bar{\partial}, \Lambda]_{gr}$ and $\bar{\partial}^* = i[\partial, \Lambda]_{gr}$, one has $[\bar{\partial}, \bar{\partial}^*]_{gr} - [\partial^*, \partial]_{gr} = i[\bar{\partial}, [\partial, \Lambda]_{gr}]_{gr} + i[[\bar{\partial}, \Lambda]_{gr}, \partial]_{gr} = i[[\bar{\partial}, \partial]_{gr}, \Lambda]_{gr} = 0.$ $\qquad\square$

We note that the above computation works to show also that $[\bar{\partial}, \bar{\partial}_h^\star]_{gr} - [\partial^\star, \partial_h]_{gr} = i[[\bar{\partial}, \partial_h]_{gr}, \Lambda]_{gr}$ if $d\omega = 0$, because $\bar{\partial}_h^\star$ and ∂_h coincide with $\bar{\partial}^\star$ and ∂ respectively at a point $x \in M$ with respect to a normal frame of E around x. The first–order terms in the Taylor expansion of the coefficients of the zero order terms of ∂_h appear in $[\bar{\partial}, \partial_h]_{gr} = \Theta_h \in C^{1,1}(M, \mathrm{Hom}(E, E))$. Identifying Θ_h naturally with an element of $\Pi_2(\mathscr{M})$, we have

Theorem 2.7 (Nakano's identity). $[\bar{\partial}, \bar{\partial}_h^\star]_{gr} - [\partial^\star, \partial_h]_{gr} = [i\Theta_h, \Lambda]_{gr}$ *if* $d\omega = 0$.

Similarly, combining Lemmas 2.3 and 2.4 we obtain:

Theorem 2.8. *For any* $\theta \in C^{0,1}(M)$, $[\bar{\partial}, \theta^*]_{gr} + [\partial^\star, \bar{\theta}]_{gr} = [i\partial\bar{\partial}\theta, \Lambda]_{gr}$ *holds if* $d\omega = 0$.

As a remark, we note that Lemma 2.3, Proposition 2.8 and Theorem 2.8 can be generalized to commutator relations for $\theta \in C^{0,1}(M, \mathrm{Hom}(E, E))$ by letting $\bar{\theta}^* = \bar{*}\bar{\theta}\bar{*}$ and $\bar{\theta}_h = h^{-1} \circ {}^t\bar{\theta} \circ h$ as follows.

Lemma 2.5. $[\theta, \Lambda] = i\bar{\theta}^*$ *for any* $\theta \in C^{0,1}(M, \mathrm{Hom}(E, E))$.

Proposition 2.11. *For any* $\sigma, \tau \in C^{0,1}(M, \mathrm{Hom}(E, E)), [\sigma, \tau^*]_{gr} + [\bar{\sigma}^*, \bar{\tau}_h]_{gr} = [[\sigma, \bar{\tau}_h]_{gr}, \Lambda]_{gr}$.

Theorem 2.9. *For any* $\theta \in C^{0,1}(M, \mathrm{Hom}(E, E))$, $[\bar{\partial}, \theta^*]_{gr} + [\partial^\star, \bar{\theta}_h]_{gr} = [i\partial\bar{\partial}\theta_h, \Lambda]_{gr}$ *holds if* $d\omega = 0$.

(M, ω) is called a **Kähler manifold** if $d\omega = 0$.

2.2.4 Positivity and L^2 Estimates

Let (M, ω) be a Kähler manifold of dimension n and let (E, h) be a Hermitian holomorphic vector bundle of rank r over M. Then, Nakano's identity implies, by integration by parts, that

$$\|\bar{\partial}u\|^2 + \|\bar{\partial}^*u\|^2 - \|\partial^*u\|^2 - \|\partial_h u\|^2 = (i(\Theta_h\Lambda - \Lambda\Theta_h)u, u)$$

holds for any $u \in C_0^{p,q}(M, E)$. In particular, one has

$$\|\bar{\partial}u\|^2 + \|\bar{\partial}^*u\|^2 \geq (i(\Theta_h\Lambda - \Lambda\Theta_h)u, u) \qquad \textbf{(basic inequality)}$$

which simplifies to

$$\|\bar{\partial}u\|^2 + \|\bar{\partial}^*u\|^2 \geq (i\Theta_h\Lambda u, u) \qquad \text{if} \quad p = n. \tag{2.17}$$

From this inequality, we shall derive a useful estimate under some positivity assumption on Θ_h which turns out to be satisfied in many situations arising in complex geometry.

Let $x \in M$ and let $(z_1, z_2, \ldots, z_n)$ be a local coordinate around x such that $\omega = \frac{i}{2} \sum dz_j \wedge d\overline{z_j}$ at x, and let $(e_1, e_2, \ldots, e_r)$ be a local frame of E. Then we define $h_{\mu\overline{\nu}} \in \mathbb{C}$ $(1 \le \mu, \nu \le r)$ by requiring $h = \sum_{\mu\nu} h_{\mu\overline{\nu}} e_\mu^* \otimes \overline{e}_\nu^*$ to hold at $x \in M$, identifying $\mathrm{Hom}(E, \overline{E}^*)$ with $E^* \otimes \overline{E}^*$. Similarly, we define $\Theta_{\alpha\overline{\beta}\nu}^{\mu} \in \mathbb{C}$ $(1 \le \alpha, \beta \le n, 1 \le \mu, \nu \le r)$ by

$$\Theta_h = \sum_{\alpha, \beta} \sum_{\mu, \nu} \Theta_{\alpha\overline{\beta}\nu}^{\mu} (e_\nu^* \otimes e_\mu) dz_\alpha \wedge d\overline{z_\beta} \tag{2.18}$$

at x, by identifying $\mathrm{Hom}(E, E)$ with $E^* \otimes E$. Then take any $u \in C^{n,q}(M, E)$ and let

$$u = \sum_J \sum_\mu u_J^\mu e_\mu dz_1 \wedge \cdots \wedge dz_n \wedge d\overline{z_J} \tag{2.19}$$

hold at x. Then $|u|^2 = 2^{n+q} \sum_J \sum_{\mu, \nu} h_{\mu\overline{\nu}} u_J^\mu \overline{u_J^\nu}$ and

$$\langle i\Theta_h \Lambda u, u \rangle = 2^{n+q} \sum_K \sum_{\alpha, \beta, \nu, \kappa} \sum_\mu \Theta_{\alpha\overline{\beta}\nu}^{\mu} h_{\mu\overline{\kappa}} u_{\{K,\alpha\}}^\nu \overline{u_{\{K,\beta\}}^\kappa} \tag{2.20}$$

at x.

Similarly,

$$\langle i\Lambda\Theta_h u, u \rangle = 2^{n+q} \sum_L \sum_{\alpha, \beta, \nu, \kappa} \sum_\mu \Theta_{\alpha\overline{\beta}\nu}^{\mu} h_{\mu\overline{\kappa}} u_{L\backslash\{\alpha\}}^\nu \overline{u_{L\backslash\{\beta\}}^\kappa} \tag{2.21}$$

at x.

Note that $\sum (\sum_\mu \Theta_{\alpha\overline{\beta}\nu}^{\mu} h_{\mu\overline{\kappa}}) \xi^{\alpha\nu} \overline{\xi^{\beta\kappa}}$ $((\xi^{\alpha\nu}) \in \mathbb{C}^{nr})$ gives a quadratic form on the fibers of $T^{1,0}M \otimes E$ as x varies.

Definition 2.1. (E, h) is said to be **Nakano positive** (resp. **Nakano semipositive**) if the quadratic form $\sum (\sum_\mu \Theta_{\alpha\overline{\beta}\nu}^{\mu} h_{\mu\overline{\kappa}}) \xi^{\alpha\nu} \overline{\xi^{\beta\kappa}}$ is positive (resp. semipositive) at every point of M. Nakano negativity and Nakano seminegativity are defined similarly.

In other words, a Hermitian holomorphic vector bundle E is said to be Nakano (semi-) positive if, for any point $x_0 \in M$, there is a neighborhood $U = U(x_0)$ with local coordinate $z = (z_1, \ldots, z_n)$ around x_0 and a coordinate $w = (w_1, \ldots, w_r)$ on the fibers of $V|_U$ coming from a holomorphic trivialization such that:

(1) over U we have the representation of the fiber metric $\sum_{\mu\overline{\nu}} h_{\mu\overline{\nu}}(z) w_\mu \overline{w_\nu}$,
(2) the matrix $(h_{\mu\overline{\nu}}(x_0))$ is the unit matrix,
(3) the total derivative $dh_{\mu\overline{\nu}}(x_0) = 0$,
    and
(4) the Hermitian form $\sum (\partial^2 h_{\mu\overline{\nu}}(x_0)/\partial z_\alpha \partial\overline{z_\beta}) \gamma_{\mu\alpha} \overline{\gamma_{\nu\beta}}$ is positive (semi-)definite.

We shall say that a holomorphic vector bundle E is Nakano (semi-)positive if it admits a fiber metric whose curvature form is (semi-)positive in the sense of Definition 2.1. In accordance with the positivity of the Kähler form ω, Nakano positivity (resp. semipositivity) of the curvature form Θ_h in the above sense will be denoted by $i\Theta_h > 0$(resp. ≥ 0). By an abuse of language we shall call the eigenvalues of Θ_h also those of $i\Theta_h$. Nakano positive (resp. semipositive) line bundles are simply called **positive** (resp. **semipositive**) line bundles. The curvature form of a positive line bundle is naturally identified with a Kähler metric.

By Corollary 2.3 and the inequality (2.17), the equality (2.20) eventually implies the following.

Theorem 2.10. *Let (M, ω) be a complete Kähler manifold and let (E, h) be a Hermitian holomorphic vector bundle over M such that $i\Theta_h - cId_E \otimes \omega \geq 0$ for some $c > 0$. Then*

$$H_{(2)}^{n,q}(M, E) = 0 \quad \text{for all } q > 0. \tag{2.22}$$

Corollary 2.4 (Kodaira–Nakano vanishing theorem). *If (M, ω) is a compact Kähler manifold and (E, h) is a Nakano positive vector bundle over M,*

$$H^{n,q}(M, E) = 0 \ (\text{or equivalently } H^q(M, \mathscr{O}(\mathbb{K}_M \otimes E)) = 0 \ ) \text{ for all } q > 0. \tag{2.23}$$

Corollary 2.5. *Positive line bundles over compact complex manifolds are ample.*

Here, a holomorphic line bundle $L \to M$ is said to be **ample** if there exists $m \in \mathbb{N}$ such that L^m is **very ample** in the sense that there exist $s_0, s_1, \ldots, s_N \in H^{0,0}(M, L^m)$ such that the ratio $(s_0 : s_1 : \cdots : s_N)$ maps M injectively to $\mathbb{CP}^{N-1}$ as a (not necessarily locally closed for noncompact M) complex submanifold.

Similarly, the inequality $\|\bar{\partial}u\|^2 + \|\bar{\partial}^*u\|^2 \geq (-i\Lambda\Theta_h u, u)$ for $u \in C_0^{0,q}(M, E)$ implies:

Theorem 2.11. *Suppose that (M, ω) is a complete Kähler manifold and there exists a $c > 0$ such that $i\Theta_h + cId_E \otimes \omega \leq 0$. Then $H_{(2)}^{0,q}(M, E) = 0$ for all $q < n$.*

Remark 2.1. The Kodaira–Nakano vanishing theorem was first established in [K-2] for line bundles. The curvature condition for vector bundles of higher rank was introduced in [N-1]. Theorem 2.10 is already sufficient for many purposes, for instance to solve the classical existence problems (cf. [Hö-1]). The reason why it works is that every holomorphic vector bundle over M is Nakano positive if M admits a strictly plurisubharmonic exhaustion function. A celebrated application of Corollary 2.4 is Kodaira's characterization of projective algebraic manifolds by the existence of positive line bundles (cf. [K-2, K-3]). The point of the following discussion is that there still remains room for quite a few refinements of Theorem 2.10 which reveal deeper truth of holomorphic functions and complex manifolds. So, instead of reviewing the well–known applications of Theorem 2.10, we shall push it a little bit further.

2.2.5 L^2 Vanishing Theorems on Complete Kähler Manifolds

Let (M, ω) be a complete Kähler manifold of dimension n and let (B, a) be a Hermitian holomorphic line bundle over M.

Theorem 2.12 (cf. [A-N] and [A-V-1]). *If $\omega = i\Theta_a$, then $H^{p,q}_{(2)}(M, B) = 0$ for $p + q > n$ and $H^{p,q}_{(2)}(M, B^*) = 0$ for $p + q < n$.*

Proof. By (2.20) and (2.21) (or by direct computation), one has

$$\langle (\omega\Lambda - \Lambda\omega)u, u \rangle = (p + q - n)|u|^2 \tag{2.24}$$

for any $u \in C^{p,q}(M, B)$, whence the conclusion follows similarly to Theorems 2.10 and 2.11. $\square$

Corollary 2.6 (Akizuki–Nakano vanishing theorem). *Let M be a compact complex manifold of dimension n and $B \to M$ a holomorphic line bundle which admits a fiber metric whose curvature form is positive. Then $H^{p,q}(M, B) = 0$ for $p + q > n$ and $H^{p,q}(M, B^*) = 0$ for $p + q < n$.*

Example 2.4. $M = \mathbb{CP}^n$, $B = \mathcal{O}(1)$ and a is the dual of the fiber metric of $\mathcal{O}(-1)$ induced by $\|\zeta\|^2$ ($\zeta \in \mathbb{C}^{n+1}$).

The following is an immediate variant of Theorem 2.12. The proof is similar.

Theorem 2.13. *If the eigenvalues $\lambda_1 \leq \ldots \leq \lambda_n$ of Θ_a with respect to ω everywhere satisfy*

$$\lambda_1 + \ldots + \lambda_p - \lambda_{q+1} - \ldots - \lambda_n \ (= \lambda_1 + \ldots + \lambda_q - \lambda_{p+1} - \ldots - \lambda_n) \geq c$$

for some positive constant c, then $H^{p,q}_{(2)}(M, B) = 0$ and $H^{n-p,n-q}_{(2)}(M, B^*) = 0$.

A refinement of Theorem 2.10 in another direction is:

Theorem 2.14 (cf. [Dm-2] and [Oh-2, Oh-8]). *Let M be a complex manifold of dimension n which admits a complete Kähler metric, and let (B, a) be a positive line bundle over M. Then, for any Kähler metric ω on M satisfying $\omega \leq i\Theta_a$, $H^{n,q}_{(2)}(M, B, \omega, a) = 0$ for $q > 0$. Moreover, for any $v \in L^{n,q}_{(2)}(M, B, \omega, a) \cap \mathrm{Ker}\bar{\partial}$, one can find $w \in L^{n,q-1}_{(2)}(M, B, \omega, a) \cap \mathrm{Dom}\bar{\partial}$ satisfying $\bar{\partial}w = v$ and $\|w\|^2 \leq q\|v\|^2$.*

Proof. Let $v \in L^{n,q}_{(2)}(M, B, \omega, a) \cap \mathrm{Ker}\bar{\partial}$. Taking any complete Kähler metric ω_∞ on M, let $\omega_\epsilon = \omega + \epsilon\omega_\infty$ for any $\epsilon \geq 0$, let $\langle , \rangle_\epsilon$ denote the pointwise inner product with respect to ω_ϵ, let Λ_ϵ denote the adjoint of $\omega_\epsilon (= \omega_\epsilon \wedge)$ with respect to ω_ϵ, let $(,)_\epsilon = \int_M \langle , \rangle_\epsilon \frac{\omega_\epsilon^n}{n!}$ and let $\| \cdot \|_\epsilon^2 = (\cdot, \cdot)_\epsilon$. Then, for any $u \in C^{n,q}_0(M, B)$,

$$\left| \int_M \langle u, v \rangle_\epsilon \frac{\omega_\epsilon^n}{n!} \right|^2 \leq (i\Theta_a \Lambda_\epsilon u, u)_\epsilon ((i\Theta_a \Lambda_\epsilon)^{-1}v, v)_\epsilon \quad \text{(Cauchy–Schwarz inequality)}.$$

$$\tag{2.25}$$

Let $x \in M$ be any point. Let $v = \sum v_J dz_1 \wedge \cdots \wedge dz_n \wedge d\overline{z_J}$, $\omega = \frac{i}{2} \sum dz_j \wedge d\overline{z_j}$ and $\omega_\infty = \frac{i}{2} \sum \theta_j dz_j \wedge d\overline{z_j}$ ($\theta_j > 0$) at x.

Then

$$\langle (i\Theta_a \Lambda_\epsilon)^{-1} v, v \rangle_\epsilon \frac{\omega_\epsilon^n}{n!} \leq \langle (\omega \Lambda_\epsilon)^{-1} v, v \rangle_\epsilon \frac{\omega_\epsilon^n}{n!}$$

$$= 2^{n+q} \sum_J \left(\sum_{j \in J} (1 + \epsilon\theta_j)^{-1} \right) |v_J|^2 \left(\prod_{j \in J} (1 + \epsilon\theta_j)^{-1} \right) \frac{\omega^n}{n!} \leq q|v|^2 \frac{\omega^n}{n!} \quad (2.26)$$

at x (for almost all x).

Hence

$$|(u, v)_\epsilon|^2 \leq q\|v\|^2 (i\Theta_a \Lambda_\epsilon u, u)_\epsilon. \quad (2.27)$$

But

$$(i\Theta_a \Lambda_\epsilon u, u)_\epsilon \leq \|\bar{\partial} u\|_\epsilon^2 + \|\bar{\partial}_\epsilon^* u\|_\epsilon^2, \quad (2.28)$$

where $\bar{\partial}_\epsilon^*$ denotes the adjoint of $\bar{\partial}$ with respect to (ω_ϵ, a).

Therefore, by Theorem 2.3, there exists for each ϵ a $w_\epsilon \in \text{Dom}\bar{\partial} \cap L_{(2)}^{p,q-1}(M, B, \omega_\epsilon, a)$ such that $\bar{\partial} w_\epsilon = v$ and

$$\|w_\epsilon\|_\epsilon^2 \leq q\|v\|^2. \quad (2.29)$$

From (2.29) one sees that there exists a locally weakly convergent subsequence of $w_{\frac{1}{k}}$ ($k \in \mathbb{N}$). The limit w satisfies $\bar{\partial} w = v$ and $\|w\|^2 \leq q\|v\|^2$. □

Since $H_{(2)}^{n,q}(M, B, i\Theta_a, a) = H_{(2)}^{n,q}(M, B, ci\Theta_a, a)$ for any $c > 0$, one has:

Corollary 2.7. *In the above situation, $H_{(2)}^{n,q}(M, B, ci\Theta_a, a) = 0$ for $q > 0$ holds for any $c > 0$.*

Corollary 2.8. *Let M be as above. Then, for any C^∞ strictly plurisubharmonic function $\Phi : M \to \mathbb{R}$,*

$$H_{(2)}^{n,q}(M, i\partial\bar{\partial}\Phi, e^{-\alpha\Phi}) = 0 \text{ for } q > 0 \quad (2.30)$$

holds for any $\alpha > 0$.

Example 2.5. $M = \mathbb{C}^n$, $\Phi = \|z\|^2$. (Compare with Example 2.2.)

Proposition 2.12. *If there exist a Kähler metric and a plurisubharmonic function Φ on M such that $\Phi^{-1}([-R, R])$ are compact for all $R \in \mathbb{R}$, then M admits a complete Kähler metric.*

Proof. Let λ be a C^∞ convex increasing function on $\mathbb{R}$ such that

$$\lambda(t) = \begin{cases} -\log(-t) & \text{if } t \leq -e^2, \\ t^2 & \text{if } t \geq 1. \end{cases}$$

Then $i\partial\bar{\partial}(\lambda \circ \Phi)$ is a complete Kähler metric on M. $\square$

If (E, h) is a Nakano positive vector bundle over a complete Kähler manifold (M, ω_∞), the above proof of Theorem 2.14 works word for word to show that $H_{(2)}^{n,q}(M, E, \omega, h) = 0$ for $q > 0$ if $d\omega = 0$ and $i\Theta_h - Id_E \otimes \omega$ is Nakano semipositive. However, if rank$E \geq 2$, the existence of such ω becomes a delicate question. Therefore, in view of applications, the following generalization of Theorem 2.14 is more appropriate.

Theorem 2.15. *Let M be a complex manifold of dimension n which admits a complete Kähler metric, and let (B, a) be a Nakano positive line bundle over M. Then, for any Kähler metric ω on M and for any $v \in L_{(2),loc}^{n,q}(M, E) \cap \text{Ker}\bar{\partial}$ satisfying $((i\Theta_a\Lambda)^{-1}v, v) < \infty$ with respect to ω, one can find $w \in L_{(2)}^{n,q-1}(M, B, \omega, a) \cap \text{Dom}\bar{\partial}$ satisfying $\bar{\partial}w = v$ and $\|w\|^2 \leq ((i\Theta_a\Lambda)^{-1}v, v)$.*

Warning. In contrast to Theorems 2.10 and 2.11, $H_{(2)}^{0,q}(M, B^*, \omega, a^*)$ may not vanish for $q < n$. For instance, $H_{(2)}^{0,0}(D, i\partial\bar{\partial}|z|^2, e^{|z|^2})$ is infinite dimensional whenever D is a (nonempty) bounded domain in $\mathbb{C}$. This suggests that the limiting procedure is essential in the proof of Theorem 2.14.

Definition 2.2. A **singular fiber metric** of a holomorphic vector bundle E over a complex manifold M is a pair (a_0, Φ) of a smooth fiber metric a_0 of E and a locally integrable function Φ on M with values in $[-\infty, +\infty)$ such that $a_0 e^{-\Phi}$ is locally equal to $\tilde{a}_0 e^{-\tilde{\Phi}}$ for some smooth fiber metric $\tilde{a}_0$ of E and some plurisubharmonic function $\tilde{\Phi}$.

The measurable section $a = a_0 e^{-\Phi}$ of $E^* \otimes \overline{E^*}$ will also be referred to as a singular fiber metric of E, and (E, a) as a singular Hermitian vector bundle. Notations such as $L_{(2)}^{p,q}(M, E, \omega, a)$ will be naturally carried over for singular Hermitian vector bundles.

Remark 2.2. The notion of singular fiber metric is naturally generalized as a certain class of measurable sections of $E^* \otimes \overline{E^*}$ which are positive definite almost everywhere (cf. [dC]).

Given a holomorphic line bundle $B \to M$ equipped with a singular fiber metric a, an ideal sheaf $\mathscr{I}_a \subset \mathscr{O} = \mathscr{O}_M$ is defined by

$$\mathscr{I}_{a,x} = \{f_x \in \mathscr{O}_x; f \in \mathscr{O}(U) \text{ and } \int_U |f|^2 e^{-\phi} dV < \infty \text{ for some neighborhood } U \ni x\},$$

(2.31)

where ϕ is a plurisubharmonic function on U such that ae^{ϕ} is a smooth fiber metric, and dV is any smooth volume form on M. $\mathscr{I}_a$ is called the **multiplier ideal sheaf** of the singular Hermitian line bundle (B, a).

For any Hermitian metric ω on M and for any open set $U \subset M$, $L_{(2),loc}^{p,q}(U, B, a)$ will denote the set of locally square integrable B-valued (p, q)-forms s on U with respect to a.

By an abuse of notation, for any d-closed $(1, 1)$-form θ on M, we shall mean by $i\Theta_a \geq i\theta$ that $\Theta_{a_0} + \partial\bar{\partial}\Phi - \theta$ is locally of the form $\partial\bar{\partial}\psi$ for some plurisubharmonic function ψ. By an abuse of language, we shall say that the singular fiber metric a has **positive curvature current** if $i\Theta_a \geq 0$ holds in the above sense.

It is remarkable that by taking a "limit" of Theorem 2.14, as Demailly did in [Dm-4], one can strengthen the assertion very much as follows.

Theorem 2.16 (cf. [Dm-3, Dm-4]). *Let M be as in Theorem 2.14, let ω be a Kähler metric on M, and let B be a holomorphic line bundle over M with a singular fiber metric a satisfying $i\Theta_a \geq \omega$. Then for any $v \in L_{(2)}^{n,q}(M, B, \omega, a) \cap \mathrm{Ker}\bar{\partial}$, one can find $w \in L_{(2)}^{n,q-1}(M, B, \omega, a) \cap \mathrm{Dom}\bar{\partial}$ satisfying $\bar{\partial}w = v$ and $\|w\|^2 \leq q\|v\|^2$.*

Proof. For any sequence of positive numbers ϵ_k converging to 0, let a_k be smooth fiber metrics of B converging to a from above and $i\Theta_{a_k} \geq (1 - \epsilon_k)\omega$. To find such a_k, first do it locally by the convolution with respect to the Kähler metric, and then patch these approximating functions together by a partition of unity. Then, by applying Theorem 2.14 for each (ω, a_k) and letting $k \to \infty$, one has the desired conclusion. $\qquad\qquad\square$

A sheaf theoretic interpretation of Theorem 2.16 has important applications (cf. Sects. 2.2.7 and 3.3.2).

Remark 2.3. The multiplier ideal sheaf was named after Kohn's work on the ideals arising in the complex boundary value problem (cf. [Kn]). Besides this, it may be worthwhile to note that $\mathscr{I}_a$ had appeared implicitly in Bombieri's work [Bb-1] which solved a question of analytic number theory by the L^2 method. In fact, Bombieri applied a theorem of Hörmander in [Hö-1] which is a prototype of Theorem 2.16.

It may be worthwhile to note that there is another limiting procedure which leads to a result of different nature. To state it, we introduce a subset $L_{(2)}^{n,q}(M, E, \sigma, h)$ of $L_{(2),loc}^{n,q}(M, E)$ (= the set of locally square integrable E-valued (n, q)-forms on M) for any smooth semipositive $(1,1)$-form σ on M as follows:

$$L_{(2)}^{n,q}(M, E, \sigma, h) = \{u; \lim_{\epsilon \to 0} \|u\|_{\epsilon} \text{ exists for any Hermitian metric } \omega_0 \text{ on } M\}.$$

$$(2.32)$$

Here $\|u\|_{\epsilon}$ denotes the norm of u with respect to $(\sigma + \epsilon\omega_0, h)$.

Note that $L_{(2)}^{n,q}(M, E, \sigma, h)$ is a Hilbert space with norm $\| \cdot \|_\sigma = \lim_{\epsilon \to 0} \| \cdot \|_\epsilon$ because of the monotonicity property "$\|u\|_\epsilon \leq \|u\|_{\epsilon'}$ if $\epsilon \geq \epsilon'$", so that $H_{(2)}^{n,q}(M, E, \sigma, h)$ and $H_{(2),loc}^{n,q}(M, E, \sigma)$ are defined similarly.

Theorem 2.17 (cf. [Oh-8, Theorem 2.8]). *Let M be a complex manifold of dimension n admitting a complete Kähler metric, and let (E, h) be a Nakano semipositive vector bundle over M. Assume that σ is a smooth semipositive $(1,1)$-form on M such that $d\sigma = 0$ and $i\Theta_h - Id_E \otimes \sigma$ is Nakano semipositive. Then $H_{(2)}^{n,q}(M, E, \sigma, h) = 0$ for $q > 0$.*

Proof. Let $v \in L_{(2)}^{n,q}(M, E, \sigma, h) \cap \mathrm{Ker}\bar{\partial}$ $(q > 0)$, let ω_0 be a complete Kähler metric on M and let $\Lambda_\epsilon = (\sigma + \epsilon\omega_0)^*$. Then

$$(((Id_E \otimes \epsilon\omega_0 + i\Theta_h)\Lambda_\epsilon)^{-1}v, v)_\epsilon \leq \|v\|_\sigma^2.$$

Hence one can find $u_\epsilon \in L^{n,q-1}(M, E, \sigma + \epsilon\omega_0, h)$ such that $\bar{\partial}u_\epsilon = v$ and $\|u_\epsilon\|_\epsilon \leq \|v\|_\sigma$. Choosing a subsequence of u_ϵ as $\epsilon \to 0$ which is locally weakly convergent, we are done. □

Remark 2.4. The prototype of Theorem 2.17 is a vanishing theorem of Grauert and Riemenschneider in [Gra-Ri-1, Gra-Ri-2] which first generalized Kodaira-Nakano's vanishing theorem for semipositive bundles. In Sects. 2.2.7 and 2.3, more on the results in this direction will be discussed.

As we have remarked in Sect. 2.2.1, $H_{(2)}^{n,q}(M)$ may not vanish even if $i\partial\bar{\partial}\Phi$ is a complete metric (e.g. $(M, \omega) = (\mathbb{C}^n, i\partial\bar{\partial}\|z\|^2)$ and $0 < q \leq n$). Nevertheless, if additionally the condition $\sup_M |d\Phi| < \infty$ is satisfied, a vanishing theorem holds for non-weighted L^2 cohomology groups.

Theorem 2.18 (Donnelly–Fefferman vanishing theorem [D-F]). *Let (M, ω) be a complete Kähler manifold of dimension n. Assume that there exists a C^∞ function $\Phi : M \to \mathbb{R}$ such that $\omega = i\partial\bar{\partial}\Phi$ and $\sup_M |d\Phi| < \infty$. Then $H_{(2)}^{p,q}(M) = 0$ for $p + q \neq n$.*

Proof. By Theorem 2.8, $[\bar{\partial}, (\bar{\partial}\Phi)^*]_{gr} + [\partial^*, \partial\Phi]_{gr} = [i\partial\bar{\partial}\Phi, \Lambda]_{gr}$. Hence, combining the assumption with (2.24) and the Cauchy–Schwarz inequality, one has

$$C(\|\bar{\partial}u\| + \|\bar{\partial}^*u\| + \|\partial^*u\| + \|\partial u\|)\|u\| \geq |p + q - n|\|u\|^2 \tag{2.33}$$

for any $u \in C_0^{p,q}(M)$. Here $C = \sup_M|\partial\Phi|$. Since

$$\|\bar{\partial}u\|^2 + \|\bar{\partial}^*u\|^2 = \|\partial u\|^2 + \|\partial u\|^2,$$

(2.33) implies an estimate equivalent to the assertion. □

Example 2.6. $M = \mathbb{B}^n$, $\Phi = -\log(1 - \|z\|^2)$ (cf. Chap. 4).

Example 2.7. $M = \mathbb{B}^n \setminus \{0\}, \Phi = -\log(-\log \|z\|^2)$ ($i\partial\bar{\partial}\Phi$ is the Poincaré metric of $\mathbb{D} \setminus \{0\}$ if $n = 1$).

We shall say that a C^∞ plurisubharmonic function Φ on M is **of self bounded gradient**, or **of SBG** for short, if

$$i(\partial\bar{\partial}\Phi - \epsilon\partial\Phi \wedge \bar{\partial}\Phi) \geq 0 \quad \text{for some } \epsilon > 0. \tag{2.34}$$

Note that, if Φ is of *SBG*, then $\arctan(\epsilon\Phi)$ is a bounded plurisubharmonic function for some $\epsilon > 0$, which is strictly plurisubharmonic if so is Φ.

With this condition of SBG, a limiting process works similarly to the proof of Theorem 2.14. For instance, one can deduce the following from Theorem 2.18.

Corollary 2.9. *Let (M, ω) be a (not necessarily complete) Kähler manifold of dimension n admitting a potential Φ of SBG. Assume that there exist an interval $I \subset \mathbb{R}$ and a C^∞ function $\lambda : I \to \mathbb{R}$ such that $\Phi(M) \subset I$, $\lambda \circ \Phi$ is of SBG and $\sup((\lambda' + \lambda'')/(\lambda')^2) < \infty$. Then $H^{p,q}_{(2)}(M, \omega) = 0$ for $p + q > n$.*

Example 2.8. $M = \{z; 0 < \|z\| < \frac{1}{e}\}$, $\Phi = \frac{-1}{\log\|z\|}$, $I = (0, 1)$, $\lambda(t) = -\log t$.

2.2.6 Pseudoconvex Cases

From the above–mentioned L^2 vanishing theorems, we shall deduce here vanishing theorems for the ordinary cohomology groups on pseudoconvex manifolds. First, vanishing theorems for positive bundles will be obtained from Theorems 2.12 and 2.14. As before, let (M, ω) be a Kähler manifold of dimension n. Recall that M is said to be **pseudoconvex** (resp. **1-complete**) if M is equipped with a C^∞ plurisubharmonic (resp. strictly plurisubharmonic) exhaustion function. Pseudoconvex manifolds are also called **weakly 1-complete** manifolds (cf. [N-2, N-3]). Since every pseudoconvex Kähler manifold admits a complete Kähler metric by Proposition 2.12, an immediate consequence of Theorem 2.12 is:

Theorem 2.19 (cf. [N-3]). *Let M be a pseudoconvex manifold of dimension n and let (B, a) be a positive line bundle over M. Then $H^{p,q}(M, B) = 0$ for $p + q > n$.*

Proof. Let $p + q > n$ and $v \in L^{p,q}_{(2),loc}(M, B) \cap \text{Ker}\bar{\partial}$. Then, it is easy to see that for any smooth plurisubharmonic exhaustion function Φ on M, one can find a convex increasing function λ such that $v \in L^{p,q}_{(2)}(M, B, i\Theta_{ae^{-\lambda\circ\Phi}}, ae^{-\lambda\circ\Phi})$. Hence, $v \in \bar{\partial}(L^{p,q-1}_{(2),loc}(M, B))$ by Theorem 2.12. $\square$

Similarly, from Theorem 2.14 we obtain:

Theorem 2.20 (cf. [Kz-1]). *Let M be a pseudoconvex manifold of dimension n and let (E, h) be a Nakano positive vector bundle over M. Then $H^{n,q}(M, E) = 0$ for $q > 0$.*

Proof. Let $q > 0$ and $v \in L^{n,q}_{(2),loc}(M, E) \cap \operatorname{Ker}\bar{\partial}$. It is clear that for the Kähler metric $i\Theta_{\det h}$ there exists a smooth plurisubharmonic exhaustion function Φ on M such that $((i\Theta_{he^{-\Phi}} \Lambda)^{-1}v, v) < \infty$ holds with respect to $(i\Theta_{\det h}, he^{-\Phi})$. Hence $v \in \operatorname{Im}\bar{\partial}$ by Theorem 2.14.   □

Similarly, but using the L^2 estimates for a sequence of solutions of the $\bar{\partial}$– equation, we have:

Theorem 2.21. *Let M be a pseudoconvex manifold of dimension n and let (E, h) be a Nakano negative vector bundle over M. Then $H^{0,q}_0(M, E) = 0$ for $q < n$.*

On the other hand, by Serre's duality theorem, Theorem 2.20 implies the following.

Theorem 2.22. *Under the situation of Theorem 2.20, $H^{0,q}_0(M, E^*) = 0$ for $q < n$.*

Remark 2.5. Nakano positivity of E is not equivalent to Nakano negativity of E^* if rank$E > 1$ (cf. [Siu-5]).

Combining Theorems 2.20 and 2.21 (or Theorem 2.22), we obtain:

Theorem 2.23. *Let M be a 1-complete manifold of dimension n. Then, for any holomorphic vector bundle E over M, $H^{0,q}(M, E) = 0$ (resp. $H^{0,q}_0(M, E) = 0$) holds for any $q > 0$ (resp. $q < n$).*

Proof. Since $H^{0,q}(M, E) \cong H^{n,q}(M, (\mathbb{K}_M)^* \otimes E)$ and $(\mathbb{K}_M)^* \otimes E$ is Nakano positive by the 1-completeness of M, that $H^{0,q}(M, E) = 0$ holds for any $q > 0$ follows from Theorem 2.20. The rest is similar.   □

Similarly to the above, one can remove the L^2 condition at infinity from Theorem 2.16. Since Theorem 2.16 can be applied for all Stein domains in M, the result can be formulated in terms of the sheaf cohomology groups as in the spirit of the theorems of de Rham and Dolbeault. We shall summarize such interpretations in the next subsection.

2.2.7   Sheaf Theoretic Interpretation

Let M be a complex manifold of dimension n and let (B, a) be a singular Hermitian line bundle over M. By $\mathscr{W}^{p,q}(B, a)$ we shall denote a sheaf over M whose sections over an open set $U \subset M$ are those elements of $L^{p,q}_{(2),loc}(U, B, a)(\subset L^{p,q}_{(2),loc}(U, B))$ whose images by $\bar{\partial}$ (in the distribution sense) belong to $L^{p,q+1}_{(2),loc}(U, B, a)$. Then, Theorem 2.16 implies that the complex $(\mathscr{W}^{n,q}(B, a)_{q\geq 0}, \bar{\partial})$ is a fine resolution of the sheaf $\mathscr{O}(\mathbb{K}_M) \otimes \mathscr{I}_a$. Hence, as an immediate consequence of Theorem 2.16 we obtain:

Theorem 2.24 (Nadel's vanishing theorem, cf. [Nd]). *In the situation of Theorem 2.16, assume moreover that M is pseudoconvex. Then*

$$H^q(M, \mathcal{O}(\mathbb{K}_M \otimes B) \otimes \mathscr{I}_a) = 0$$

for $q > 0$.

Theorem 2.24 was proved by Nadel [Nd] for the compact case who applied it to prove the existence of Kähler–Einstein metrics on certain projective algebraic manifolds. Nadel proved also the coherence of $\mathscr{I}_a$ (cf. Chap. 3).

As well as the generalization to noncompact manifolds, Nadel's vanishing theorem can be extended easily to complex spaces with singularities.

Definition 2.3. Given a reduced complex space X of pure dimension n, a sheaf ω_X over X is called the L^2-**dualizing sheaf** of X if

$$\Gamma(U, \omega_X) = \{u \in \Gamma(U \cap X_{reg}, \mathcal{O}(\mathbb{K}_{X_{reg}})); u \in L^{n,0}_{(2)}(V \cap X_{reg}) \text{ for every } V \Subset U\}$$

for any open subset $U \subset X$.

Similarly, for any singular Hermitian line bundle (B, a) over X, the **multiplier dualizing sheaf** $\omega_{X,a}$ is defined as the collection of square integrable germs of B-valued holomorphic n-forms with respect to a.

For any desingularization $\pi : \tilde{X} \to X$, one has $\omega_X = \pi_* \mathcal{O}(\mathbb{K}_{\tilde{X}})$, since the support of effective divisors are negligible as the singularities of L^2 holomorphic functions (cf. Theorem 1.8). In particular, ω_X is a coherent analytic sheaf over X.

Theorem 2.25 (Nadel's vanishing theorem on complex spaces). *Let X be a reduced and pseudoconvex complex space of pure dimension n with a Kähler metric, and let (B, a) be a singular Hermitian line bundle over X. Then*

$$H^q(X, \mathcal{O}(B) \otimes \omega_{X,a}) = 0$$

for $q > 0$.

Generally speaking, precise vanishing theorems are important in complex geometry not only because they yield effective results, but also because they give a wider perspective in the theory of symmetry and invariants. In fact, their generalizations and variants have been found in the literature of geometry, analysis and algebra. Let us review an example of such a development motivated by more algebraic ideas.

Soon after the appearance of [Gra-Ri-1, Gra-Ri-2], Ramanujam [Rm-1] came up with a similar generalization of Kodaira's vanishing theorem. Ramanujam's background was Grothendieck's theory [Grt-1, Grt-2], which is a foundation of algebraic geometry over the fields of arbitrary characteristic. In this situation, he could prove a vanishing theorem only for surfaces. No one could have done better because it is false for higher dimensions in positive characteristic. Note that a counterexample to Kodaira's vanishing theorem in positive characteristic was found only after the publication of [Rm-2] (cf. [Rn]). Anyway, from this new perspective, inspired also by Bombieri's work [Bb-2] on pluricanonical surfaces, he could strengthen the vanishing theorem in the following way.

Theorem 2.26 (cf. [Rm-2]). *Let X be a nonsingular projective algebraic variety of dimension* $n \geq 2$ *over* $\mathbb{C}$ *and let* $L \to X$ *be a holomorphic line bundle whose first Chern class* $c_1(L)$ *satisfies* $c_1(L)^2 > 0$ *and* $c_1(L) \cdot \mathscr{C} (:= \deg L|_{\mathscr{C}_{red}}) \geq 0$ *for any compact complex curve* $\mathscr{C}$ *in X. Then* $H^1(X, \mathscr{O}(L^*)) = 0$.

In the case where $L = [D]$ for some effective divisor D on X, $H^1(X, \mathscr{O}(L^*)) = 0$ implies in particular that the support $|D|$ of D is connected.

The formulation of Ramanujam's theorem is in the same spirit as in Nakai's numerical criterion for ampleness of line bundles (cf. [Na-1, Na-2, Na-3]). Mumford [Mm] gave an alternate proof of Theorem 2.26 but did not proceed to the higher–dimensional cases. Kawamata [Km-1] and Viehweg [V] overcame this shortcoming independently by establishing the following:

Theorem 2.27. *Let X be a nonsingular projective algebraic variety of dimension* n *and let* $L \to X$ *be a holomorphic line bundle with* $c_1(L)^n > 0$ *such that* $c_1(L) \cdot \mathscr{C} \geq 0$ *for any compact complex curve* $\mathscr{C}$ *in X. Then*

$$H^k(X, \mathscr{O}(\mathbb{K}_X \otimes L)) = 0$$

holds for all $k > 0$.

By the Serre duality, this contains Theorem 2.26 as a special case. Being a numerical criterion for the cohomology vanishing, Theorem 2.27 is of basic importance in birational geometry. (See [Km-M-M] for instance.) A simple analytic proof of Theorem 2.27 was later given by Demailly [Dm-4]. According to the modern terminology, L is said to be **nef** (= numerically effective or numerically eventually free) if $c_1(L) \cdot \mathscr{C} \geq 0$ for any complex curve $\mathscr{C}$ on X. The notion of nef line bundles naturally extends to Kähler manifolds. Namely, a holomorphic line bundle L over a compact Kähler manifold is called **nef** if $c_1(L)$ is in the closure of the cone of Kähler classes. Nef bundles over nonsingular projective varieties are nef in the latter sense because any Kähler class is in the closure of the cone generated over $\mathbb{R}_+$ by the first Chern classes of positive line bundles. Nef line bundles also make sense over the proper images of compact Kähler manifolds by almost biholomorphic maps. Furthermore, in virtue of a theorem of Varouchas [Va], they can be defined similarly over proper holomorphic images of compact Kähler manifolds. Demailly and Peternell [Dm-P] proved:

Theorem 2.28. *Let X be a compact and normal complex space of dimension* n *admitting a Kähler metric, and let* L *be a nef line bundle over X with* $c_1(L)^2 \neq 0$. *Then* $H^q(X, \mathscr{O}(\mathbb{K}_X \otimes L)) = 0$ *for* $q \geq n - 1$.

Similarly to Theorem 2.26, if $n \geq 2$ and $L = [D]$ for some effective divisor D, it follows from the assumption of Theorem 2.28 that $|D|$ is connected. In [Oh-27] it is proved that $H_0^1(X \setminus |D|, \mathscr{O}) = 0$ under the same hypothesis. Recall that H_0^1 denotes the cohomology with compact support. A question was raised in [Dm-P-S] whether or not $H^q(X, \mathscr{O}(\mathbb{K}_X \otimes L)) = 0$ holds if L is nef and $c_1(L)^{n-q+1} \neq 0$. It was recently settled by J.-Y.Cao [CJ-1].

There exists another sheaf theoretic interpretation of Theorem 2.14, slightly different from that of Theorem 2.16. It is based on an observation that, although $L_{(2)}^{n,q}(M, E, \sigma, h)$ are subsets of $L_{(2),loc}^{n,q}(M, E)$ for semipositive σ, $H_{(2),loc}^{n,q}(M, E, \sigma)$ can be naturally isomorphic to certain sheaf cohomology group on a quotient space of M. Assume that there exists a proper holomorphic map π from M to a complex space X admitting a Kähler metric ω. Then, Theorem 2.14 implies that $H_{(2),loc}^{n,q}(M, E, \pi^*\omega) \cong H^q(X, \pi_*\mathcal{O}(\mathbb{K}_M \otimes E))$ for $q \geq 0$ if E is Nakano semipositive on some neighborhood of $\pi^{-1}(x)$ for any $x \in X$. In particular, one has the following.

Theorem 2.29 (cf.[Oh-8, Theorem 3.1]). *Let M be a pseudoconvex Kähler manifold, let X be a complex space with a Kähler metric ω, and let $\pi : M \to X$ be a holomorphic map. Then, for any Nakano semipositive vector bundle (E, h) over M satisfying $i\Theta_h - Id_E \otimes \pi^*\omega \geq 0$, $H^q(X, \pi_*\mathcal{O}(\mathbb{K}_M \otimes E)) = 0$ for $q > 0$.*

Remark 2.6. In [Oh-8], this was stated only for compact M. As in the case of Theorem 2.25, Theorem 2.29 has several predecessors besides the Kodaira-Nakano vanishing theorem. These are semipositivity theorems for the direct image sheaves of the relative canonical bundles $\mathbb{K}_M \otimes \pi^*\mathbb{K}_X^{-1}$ for Kählerian M and nonsingular X (cf. [Gri-1, Gri-3, F-1]). Quite recently, there was an unexpected development in the theory of L^2 holomorphic functions closely related to the semipositivity theorem of this type (see Chaps. 3 and 4).

Roughly speaking, the choice of singular fiber metrics and degenerate base metrics (= pseudometrics) amounts to the choice of boundary conditions in the problems of partial differential equations. There are inductive arguments to produce "good" singular fiber metrics (cf. Lemma 3.2 in Chap. 3).

Concerning the L^2 $\bar{\partial}$-cohomology groups of type (p, q), there are results which relate analytic invariants and topological invariants on complex spaces with singularities. Let us review some of these results in the next subsection.

2.2.8 Application to the Cohomology of Complex Spaces

The symmetry $[\bar{\partial}, \bar{\partial}^*]_{gr} = [\partial^*, \partial]_{gr}$ of the complex Laplacian on a compact Kähler manifold M yields Hodge's decomposition theorem

$$H^r(M) \cong \oplus_{p+q=r} H^{p,q}(M)$$

$$H^{p,q}(M) \cong \overline{H^{q,p}(M)}$$

(cf. [W]). Hence it is natural to expect that the corresponding decomposition for the spaces of L^2 harmonic forms carries similar geometric information also on some noncompact Kähler manifolds. Grauert [Gra-2] has shown that, for every compact Kähler space X, X_{reg} carries a complete Kähler metric. Based on this, it will be shown below after [Sap] and [Oh-10, Oh-11, Oh-14] that the L^2 cohomology groups of X_{reg} with respect to some class of (not necessarily complete) metrics are canonically isomorphic to the ordinary ones in certain degrees. Accordingly, the Hodge decomposition remains true for $M = X_{reg}$ there. The method also gives a partial solution to a conjecture of Cheeger, Goresky and MacPherson in [C-G-M]. Related results are also reviewed.

Let X be a (reduced) complex space. A **Hermitian metric** on X will be defined as a Hermitian metric or a positive $(1,1)$-form say ω on X_{reg} such that, for any point $x_0 \in$ Sing$X$ there exists a neighborhood $U \ni x_0$ in X, a proper holomorphic embedding ι of U into a polydisc $\mathbb{D}^N$ for some N and a C^∞ positive $(1,1)$ form Ω on $\mathbb{D}^N$ such that $\omega = \iota^* \Omega$ holds on $U \cap X_{reg}$. A complex space X equipped with a Hermitian metric will be called a **Hermitian complex space**. A Hermitian complex space (X, ω) is called a **Kähler space** if $d\omega = 0$.

Lemma 2.6. *Let (X, ω) be a compact Kähler space. Then there exists a continuous function $\varphi : X \to [0, 1]$ such that $\omega + i\partial\bar{\partial}\varphi$ is a complete Kähler metric on X_{reg} for which the length of $\partial\varphi$ is bounded.*

Proof. Let $x_0 \in$ Sing$X$ be any point and let $f_1, \ldots, f_m$ be holomorphic functions on a neighborhood U of x_0 which generate the ideal sheaf $\mathscr{I}_{\mathrm{Sing}X}$ on U. Then, by shrinking U if necessary so that $\sum_{j=1}^{m} |f_j|^2 < e^{-e}$ on U, we put

$$\varphi_U = \frac{1}{\log\left(-\log \sum_{j=1}^{m} |f_j|^2\right)}.$$

By an abuse of notation, we put $\varphi_U = 1$ if $U \cap$ Sing$X = \varnothing$. In order to see that a desired function φ can be obtained by patching φ_U by a partition of unity, let us take another generator $(g_1, \ldots, g_\ell)$ of $\mathscr{I}_{\mathrm{Sing}X}$ over U. Then it is easy to verify that one can find a neighborhood $V \ni x_0$ such that $\omega + i\partial\bar{\partial}\varphi_U$ and $\omega + i\partial\bar{\partial}(\log(-\log \sum_{k=1}^{\ell} |g_k|^2))^{-1}$ are positive and quasi-isometrically equivalent to each other on $X_{reg} \cap V$. Moreover, it is also immediate that, for any $\epsilon > 0$ one can find a neighborhood $W \ni x_0$ such that the length of $\partial\varphi_U$ on $W \cap X_{reg}$ with respect to $\omega + i\partial\bar{\partial}\varphi_U$ (or even with respect to $i\partial\bar{\partial}\varphi_U$) is less than ϵ.

Now let $\mathscr{U} = \{U_\alpha\}_{\alpha \in A}$ be a finite open cover of X by such U, and let ρ_α be a C^∞ partition of unity associated to $\mathscr{U}$. We put

$$\varphi = \sum_{\alpha \in A} \rho_\alpha \varphi_{U_\alpha}. \tag{2.35}$$

Then φ is a continuous function on X with values in $[0, 1]$ such that $\varphi^{-1}(0) = $ Sing$X$ and $\varphi|_{X_{reg}}$ is C^∞. Furthermore, since $\partial(\sum \rho_\alpha) = 0$ and $\partial\bar{\partial}(\sum \rho_\alpha) = 0$,

$$\partial\bar{\partial}\varphi = \sum \partial\bar{\partial}\rho_\alpha(\varphi_{U_\alpha} - \varphi_{U_\beta}) + \sum \partial\rho_\alpha(\bar{\partial}\varphi_{U_\alpha} - \bar{\partial}\varphi_{U_\beta})$$

$$+ \sum (\partial\varphi_{U_\alpha} - \partial\varphi_{U_\beta})\bar{\partial}\rho_\alpha + \sum \rho_\alpha \partial\bar{\partial}\varphi_{U_\alpha}. \tag{2.36}$$

Combining (2.34) with the above remarks on $\omega + i\partial\bar{\partial}\varphi_U$ and $\partial\varphi_U$, it is clear that $\omega + \epsilon i\partial\bar{\partial}\varphi$ is a complete Kähler metric on X_{reg} for $0 < \epsilon \ll 1$. $\qquad\square$

Any function φ of the form (2.35) will be called a **Grauert potential** on X. A metric of the form $\omega + i\partial\bar{\partial}\varphi$ will then be called a **Grauert metric**. The boundedness condition for $\partial\varphi$ is important when one wants to extend the Hodge theory to complex spaces with singularities. A basic fact for that is the following:

Proposition 2.13. *Let (M, ω) be a complete Kähler manifold, let (E, h) be a Hermitian holomorphic vector bundle over M and let φ be a real-valued bounded C^∞ function on M such that $\partial\varphi$ is bounded with respect to ω. Then, for any nonnegative integers p and q, and for any $u \in L^{p,q}_{(2)}(M.E)$ satisfying $u \in \text{Dom}\bar{\partial} \cap \text{Dom}\bar{\partial}^*$, u belongs to the domain of the adjoint of $\bar{\partial}$ with respect to the modified fiber metric $he^{-\varphi}$. If moreover Θ_h and $\partial\bar{\partial}\varphi$ are also bounded, u belongs also to the domains of ∂_h, $\partial_{he^{-\varphi}}$ and ∂^*.*

Proof. The proof of the first assertion follows immediately from the definition of the adjoint of $\bar{\partial}$. The second assertion follows from Nakano's identity. $\qquad\square$

For any open set $U \subset X$, the L^2 cohomology groups $H^{p,q}_{(2)}(U \cap X_{reg}), H^r_{(2)}(U \cap X_{reg})$ with respect to ω will be denoted by $H^{p,q}_{(2)}(U), H^r_{(2)}(U)$, for simplicity. The L^2 cohomology groups with supports restricted to relatively compact subsets of U will be denoted by $H^{p,q}_{(2),0}(U), H^r_{(2),0}(U)$. Similarly, the L^2 cohomology groups with respect to a Grauert metric $\omega + i\partial\bar{\partial}\varphi$ and those with "compact support in U" will be denoted by $H^{p,q}_{(2),\varphi}(U), H^r_{(2),\varphi}(U), H^{p,q}_{(2),\varphi,0}(U)$ and $H^r_{(2),\varphi,0}(U)$. Then the vanishing of the L^2 cohomology of Akizuki–Nakano type on complete Kähler manifolds implies that these L^2 cohomology groups do not see the singularities in higher degrees. For instance the following holds.

Theorem 2.30. *Let (X, ω) be a compact Kähler space of pure dimension n and let φ be a Grauert potential on X. If $\dim \text{Sing} X = 0$, then*

$$H^{p,q}_{(2)}(X) \cong H^{p,q}_{(2),\varphi}(X) \cong H^{p,q}_0(X_{reg}) \quad \text{for } p + q > n + 1$$

and

$$H^r_{(2)}(X) \cong H^r_{(2),\varphi}(X) \cong H^r_0(X_{reg}) \quad \text{for } r > n + 1.$$

Moreover, the natural homomorphisms from $H^{p,q}_0(X_{reg})$ to $H^{p,q}_{(2)}(X)$ and $H^{p,q}_{(2),\varphi}(X)$ are surjective for $p + q = n + 1$, and so are those from $H^{n+1}_0(X_{reg})$ to $H^{n+1}_{(2)}(X)$ and $H^{n+1}_{(2),\varphi}(X)$.

Proof. Let $x_0 \in \text{Sing} X$ and let $W \ni \{x_0\}$ be a neighborhood such that $W \cap \text{Sing} X = \{x_0\}$. Let $\delta > 0$ be sufficiently small so that $W_\delta := \{x \in W; \varphi(x) < \delta\} \Subset W$. Then, with respect to the complete Kähler metric $\omega_{\epsilon,\delta} := \omega + \epsilon i\partial\bar{\partial}(\log(\delta - \varphi))^{-1}$ $(0 < \epsilon \ll \delta < 1)$ on $W_\delta \setminus \{x_0\}$, the L^2 cohomology groups $H^{p,q}_{(2)}(W_\delta \setminus \{x_0\}, \omega_{\epsilon,\delta})$

vanish for $p + q > n$ by Theorem 2.12. On the other hand, it is easy to see that, for any L^2 (p,q) form f on W_{reg} with $p + q > n$, $f|_{W_{reg}\setminus\{x_0\}}$ is L^2 with respect to $\omega_{\epsilon,\delta}$. Therefore the natural homomorphism

$$H_0^{p,q}(X_{reg}) \to H_{(2),\varphi}^{p,q}(X)$$

is surjective if $p + q > n$ and injective if $p + q > n + 1$. Concerning the L^2 de Rham cohomology groups $H_{(2)}^r$, that $H_{(2)}^r(W_\delta \setminus \{x_0\}, \omega_{\epsilon,\delta})$ vanish for $r > n$ can be shown as follows: Let $u \in L_{(2)}^r(W_\delta \setminus \{x_0\}, \omega_{\epsilon,\delta})$, $du = 0$ and $r > n$. Then u is decomposed as $u = u^{r,0} + u^{r-1,1} + \cdots + u^{0,r}$ with $u^{p,q} \in L_{(2)}^{p,q}(W_\delta \setminus \{x_0\}, \omega_{\epsilon,\delta})$. Since $du = 0$, $\bar{\partial}u^{0,r} = 0$, so that there exists $v \in L_{(2)}^{0,r-1}(W_\delta \setminus \{x_0\}, \omega_{\epsilon,\delta})$ such that $\bar{\partial}v = u^{0,r}$ and $\bar{\partial}^*v = 0$. By Proposition 2.13, it follows in particular that $\partial v \in L_{(2)}^{1,r-1}(W_\delta\setminus\{x_0\}, \omega_{\epsilon,\delta})$. Hence $u - dv \in \oplus_{j=1}^r L_{(2)}^{j,r-j}(W_\delta \setminus \{x_0\}, \omega_{\epsilon,\delta})$. Proceeding similarly, we obtain that the L^2 de Rham cohomology class of u is 0. Thus we obtain the assertion for $H_{(2),\varphi}^{p,q}(X)$ and $H_{(2),\varphi}^r(X)$. As for the ordinary L^2 cohomology groups $H_{(2)}^{p,q}(X)$ and $H_{(2)}^r(X)$, they are considered respectively as the limits of $H_{(2),\varphi}^{p,q}(X)$ and $H_{(2),\varphi}^r(X)$. For that, we make a special choice of φ as in the proof of Lemma 2.6. Then, after fixing δ, we consider ω as the limit of $\omega_{\epsilon,\delta}$ ($\epsilon \to 0$). Then, as is easily checked, $L_{(2)}^{p,q}(W_\delta\setminus\{x_0\}, \omega) \subset L_{(2)}^{p,q}(W_\delta\setminus\{x_0\}, \omega_{\epsilon,\delta})$ if $p+q > n$, so that by solving the $\bar{\partial}$ equation $\bar{\partial}v_\epsilon = u$ with L^2 norm estimates for $v_\epsilon \in L_{(2)}^{p,q-1}(W_\delta\setminus\{x_0\}, \omega_{\epsilon,\delta})$ uniformly in ϵ, and by taking the weak limit of a subsequence of v_ϵ, we obtain the required the vanishing results for $H_{(2)}^{p,q}(W_\delta \setminus \{x_0\}, \omega)$ as well as those for $H_{(2)}^r(W_\delta \setminus \{x_0\}, \omega)$. $\quad\square$

In view of the long exact sequences

$$\cdots \to H_0^{p,q}(X_{reg}) \to H_{(2)}^{p,q}(X) \to \varinjlim H_{(2)}^{p,q}(X \setminus K) \to H_0^{p,q+1}(X_{reg}) \to \cdots$$

(resp. $\cdots \to H_0^{p,q}(X_{reg}) \to H_{(2),\varphi}^{p,q}(X) \to \varinjlim H_{(2),\varphi}^{p,q}(X \setminus K) \to H_0^{p,q+1}(X_{reg}) \to \cdots$),

where $\varinjlim$ denotes the inductive limit of the system

$$H_{(2)}^{p,q}(X \setminus K_1) \to H_{(2)}^{p,q}(X \setminus K_2)$$

(resp. $H_{(2),\varphi}^{p,q}(X \setminus K_1) \to H_{(2),\varphi}^{p,q}(X \setminus K_2)$) $(K_1 \subset K_2 \Subset X_{reg})$,

Theorem 2.30 says

$$\varinjlim H_{(2)}^{p,q}(X \setminus K) = \varinjlim H_{(2),\varphi}^{p,q}(X \setminus K) = 0 \text{ if } p+q > n. \tag{2.37}$$

The proof shows that

$$\varinjlim H_{(2)}^r(X \setminus K) = \varinjlim H_{(2),\varphi}^r(X \setminus K) = 0 \text{ for } r > n$$

is a consequence of (2.37). Existence of the natural homomorphisms

$$\lim_{\to} H^{p,q}_{(2),\varphi}(X \setminus K) = \lim_{\to} H^{p,q}_{(2)}(X \setminus K), \quad p+q > n$$

is crucial to deduce

$$\lim_{\to} H^{p,q}_{(2)}(X \setminus K) = 0$$

from

$$\lim_{\to} H^{p,q}_{(2),\varphi}(X \setminus K) = 0.$$

The Kähler condition is superfluous here. Obviously Theorem 2.30 holds for compact Hermitian complex spaces. Thus, an essential part of Theorem 2.30 can be stated as a Dolbeaut–type lemma:

Lemma 2.7. *Let V be an analytic set of pure dimension n in $\mathbb{D}^N$ containing z_0 as an isolated singularity. Then there exists a neighborhood $U \ni z_0$ such that $H^{p,q}_{(2)}(U) = 0$, $H^{p,q}_{(2),\varphi}(U) = 0$ for $p + q > n$ and $H^r_{(2)}(U) = 0$, $H^r_{(2),\varphi}(U) = 0$ for $r > n$.*

Because of the presence of singularities, it is not allowed immediately to apply the ordinary duality theorems due to Poincaré and Serre to obtain the results for $p + q$, $r < n$, simply reversing the direction of the arrows. Nevertheless there is a method to prove the following (see [Oh-10, Supplement]).

Lemma 2.8. *Let V and z_0 be as above. Then there exists a neighborhood $U \ni z_0$ such that*

$$H^{p,q}_{(2),0}(U) = 0, \quad H^{p,q}_{(2),\varphi,0}(U) = 0 \text{ for } p + q < n,$$

and

$$H^r_{(2),0}(U) = 0, \quad H^r_{(2),\varphi,0}(U) = 0 \text{ for } r < n.$$

As a result, the dual of Theorem 2.30 is stated as follows.

Theorem 2.31. *Let (X, ω) and φ be as in Theorem 2.30. Then*

$$H^{p,q}_{(2)}(X) \cong H^{p,q}_{(2),\varphi}(X) \cong H^{p,q}(X_{reg}) \quad \text{for } p + q < n - 1$$

and

$$H^r_{(2)}(X) \cong H^r_{(2),\varphi}(X) \cong H^r_0(X_{reg}) \quad \text{for } r < n - 1.$$

Moreover, the natural homomorphisms from $H^{p,q}_{(2)}(X)$ and $H^{p,q}_{(2),\varphi}(X)$ to $H^{p,q}(X_{reg})$ are injective for $p + q = n - 1$, and so are those from $H^{n-1}_{(2)}(X)$ and $H^{n-1}_{(2),\varphi}(X)$ to $H^{n-1}(X_{reg})$.

Proof. First we shall show the surjectivity of

$$H^{p,q}_{(2),\varphi}(X) \to H^{p,q}(X_{reg}) \quad \text{for } p+q < n-1.$$

For that, it suffices to prove that, for any $u \in C^{p,q}(X_{reg}) \cap \mathrm{Ker}\bar\partial$ $(p+q < n-1)$, there exists $w \in L^{p,q-1}_{(2),loc}(X_{reg})$ such that $u - \bar\partial w \in L^{p,q}_{(2),\varphi}(X)$. Let ρ be a C^∞ function on X such that $\rho = 1$ on a neighborhood of SingX and supp$\rho \subset \{\varphi < \delta\}$ for sufficiently small δ. We put $V_\delta = \{\varphi < \delta\} \setminus \mathrm{Sing}X$. Then take any $v \in L^{p,q}_{(2)}(V_\delta, i\partial\bar\partial(\frac{1}{\log(\delta-\varphi)}))$ with $\bar\partial v = \bar\partial(\rho u)$ on V_δ. (Note that $\omega + i\partial\bar\partial\varphi + i\partial\bar\partial(\frac{1}{\log(\delta-\varphi)})$ is a complete Kähler metric on V_δ.) We put

$$\tilde{v} = \begin{cases} v & on \ V_\delta \\ 0 & on \ X_{reg} \setminus V_\delta. \end{cases}$$

Then $\tilde{v} \in L^{p,q}_{(2),\varphi}(X_{reg})$, $\bar\partial(\tilde{v} - \rho u) = 0$ and supp$(\tilde{v} - \rho u) \subset V_\delta$. Hence, similarly to the above, by applying Theorem 2.13 for $a = e^{-\mu(\varphi)}$ for a family of convex increasing functions μ, one can find $w \in L^{p,q-1}_{(2),loc}(X_{reg})$ such that supp$w \subset V_\delta$ and $\tilde{v} - \rho u = \bar\partial w$. Hence $u - \bar\partial w \in L^{p,q}_{(2),\varphi}(X)$. Considering a long exact sequence, we conclude that the natural homomorphisms

$$H^{p,q}_{(2),\varphi}(X) \to H^{p,q}(X_{reg})$$

are bijective if $p+q < n-1$ and injective if $p+q = n-1$. Moreover, since there are natural inclusions $L^{p,q}_{(2)}(X_{reg}, \omega + i\partial\bar\partial\varphi) \subset L^{p,q}_{(2)}(X_{reg}, \omega)$ for $p+q < n$, surjectivity of the induced homomorphisms $H^{p,q}_{(2),\varphi}(X) \to H^{p,q}_{(2)}(X)$ for $p+q < n-1$ follows similarly to the above in view of Lemma 2.8.   □

Theorems 2.30 and 2.31 can be easily generalized to compact Hermitian complex spaces with isolated singularities, and naturally extended to the spaces with arbitrary singularities. In the latter case, the effective ranges of bijectivity between the L^2 and the ordinary cohomology groups become narrower (cf. [Oh-11]).

By restricting ourselves only to the L^2 de Rham cohomology, we have the following partial answer to a conjecture of Cheeger, Goresky and MacPherson [C-G-M] on the equivalence between the L^2 cohomology and the intersection cohomology (of the middle perversity) for compact complex spaces.

Theorem 2.32 (cf. [Oh-14, Oh-15]). *Let (X, ω) be a compact Hermitian complex space of pure dimension n such that* dim SingX $= 0$. *Then*

$$H^r_{(2)}(X) \cong H^r(X_{reg}) \quad \text{for } r < n$$

$$H^n_{(2)}(X) \cong \mathrm{Im}(H^n_0(X_{reg}) \to H^n(X_{reg}))$$

and

$$H^r_{(2)}(X) \cong H^r_0(X_{reg}) \quad \text{for } r > n.$$

Sketch of Proof. In [Sap], Saper proved that there exists a complete metric $\tilde{\omega}$ on X_{reg} such that

$$H_{(2)}^r(X_{reg}, \tilde{\omega}) \cong H^r(X_{reg}) \quad for \ r < n$$

$$H_{(2)}^n(X_{reg}, \tilde{\omega}) \cong \mathrm{Im}(H_0^n(X_{reg}) \to H^n(X_{reg}))$$

and

$$H_{(2)}^r(X_{reg}, \tilde{\omega}) \cong H_0^r(X_{reg}) \quad for \ r > n$$

hold. By analyzing the behavior of $H_{(2)}^r(X_{reg}, \omega + \epsilon\tilde{\omega})$ and $H_{(2)}^{p,q}(X_{reg}, \omega + \epsilon\tilde{\omega})$ as $\epsilon \to 0$, the required isomorphisms are obtained. $\square$

Since Saper's metric is Kählerian if so is ω, Theorem 2.32 naturally implies an extension of Hodge's decomposition theorem to compact Kähler spaces with isolated singularities.

On the other hand, Theorems 2.30 and 2.31 with the Kähler condition implies the following:

Theorem 2.33. *Let* (X, ω) *be a compact Kähler space of pure dimension* n *with* $\dim \mathrm{Sing} X = 0$. *Then*

$$H_0^r(X_{reg}) \cong \oplus_{p+q=r} H_0^{p,q}(X_{reg}) \quad for \ r > n+1$$

$$(resp. \ H^r(X_{reg}) \cong \oplus_{p+q=r} H^{p,q}(X_{reg}) \quad for \ r < n-1)$$

and

$$H_0^{p,q}(X_{reg}) \cong \overline{H_0^{q,p}(X_{reg})} \quad for \ p+q > n+1$$

$$(resp. \ H^{p,q}(X_{reg}) \cong \overline{H^{q,p}(X_{reg})} \quad for \ p+q < n-1).$$

Since $\dim H_0^r(X_{reg}) < \infty$ and $\dim H^r(X_{reg}) < \infty$ for all r, the following is immediate from Theorem 2.33.

Theorem 2.34. *In the situation of Theorem 2.33,* $\dim H_0^{p,q}(X_{reg}) < \infty$ *for* $p+q > n+1$ *and* $\dim H^{p,q}(X_{reg}) < \infty$ *for* $p+q < n-1$.

Of course the above–mentioned proof of Theorem 2.34 collapses at once if the Kählerianity assumption is removed. Nevertheless the finite dimensionality conclusion itself remains true. More generally the following is true:

Theorem 2.35 (cf.[Oh-11]). *Let* X *be a compact complex space of pure dimension* n *with* $\dim \mathrm{Sing} X = k$. *Then*

$$\dim H_0^{p,q}(X_{reg}) < \infty \quad for \ p+q > n+1+k$$

and

$$\dim H^{p,q}(X_{reg}) < \infty \quad for \ p+q < n-1-k.$$

Actually Theorem 2.35 is a special case of a more general finiteness theorem due to Andreotti and Grauert in [A-G]. The L^2 approaches towards it will be reviewed in the next section.

Remark 2.7. Complete Kähler metrics naturally live on locally symmetric varieties, and the L^2 cohomology is known to be isomorphic to the intersection cohomology there (cf. [Sap-St]). For basic theorems and the background on the L^2 cohomology of such a distinguished class of metrics, see [Z], [K-K] and [Fj-2].

2.3   Finiteness Theorems

Given a complex manifold M and a holomorphic vector bundle E over M, we have seen that certain $L^2 \bar{\partial}$-cohomology groups $H^{p,q}_{(2)}(M, E)$ vanish under some conditions on the metrics of M and E. Following a basic argument in [Hö-1, Theorem 3.4.1 and Lemma 3.4.2], we are going to see below that, by throwing away the positivity assumption on the curvature form of E, but only on a compact subset of M, one has finite dimensionality of the $L^2 \bar{\partial}$-cohomology instead of its vanishing. To derive the finite–dimensionality as well for the ordinary $\bar{\partial}$-cohomology, a limiting procedure is applied which is reminiscent of Runge's approximation theorem.

2.3.1 L^2 Finiteness Theorems on Complete Manifolds

Let (M, ω) be a complete Hermitian manifold of dimension n and let (E, h) be a Hermitian holomorphic vector bundle over M. If there exists a compact set $K_0 \subset M$ such that $d\omega = 0$ holds on $M \setminus K_0$, the basic inequality in Sect. 2.2.4 implies that

$$\|\bar{\partial}u\|^2 + \|\bar{\partial}^*u\|^2 \geq (i(\Theta_h\Lambda - \Lambda\Theta_h)u, u) \qquad (2.38)$$

holds for any $u \in C^{p,q}_0(M \setminus K_0, E)$.

Therefore, for any neighborhood $U \supset K_0$, one can find a constant $C_U > 0$ such that

$$(1 + \epsilon)(\|\bar{\partial}u\|^2 + \|\bar{\partial}^*u\|^2) + \frac{C_U}{\epsilon}\int_U |u|^2\frac{\omega^n}{n!} \geq (i(\Theta_h\Lambda - \Lambda\Theta_h)u, u) \qquad (2.39)$$

holds for any $\epsilon > 0$ and $u \in C^{p,q}_0(M, E)$.

Recalling a basic part of real analysis, by the strong ellipticity of the differential operator $\bar{\partial}\bar{\partial}^* + \bar{\partial}^*\bar{\partial}$ and Rellich's lemma, one has:

Lemma 2.9. *For any compact subset K of M, any sequence $u_k \in \mathrm{Dom}\bar{\partial} \cap \mathrm{Dom}\bar{\partial}^* \cap L_{(2)}^{p,q}(M, E)$ satisfying*

$$\sup \left(\|u_k\| + \|\bar{\partial}u_k\| + \|\bar{\partial}^* u_k\| \right) < \infty$$

admits a subsequence u_{k_μ} ($\mu \in \mathbb{N}$) such that

$$\lim_{\mu,\nu\to\infty} \int_K |u_{k_\mu} - u_{k_\nu}|^2 \frac{\omega^n}{n!} = 0.$$

Hence, in view of Proposition 2.2, we are naturally led to the following finiteness theorem.

Theorem 2.36. *Suppose that $i\Theta_h - c\mathrm{Id}_E \otimes \omega > 0$ holds for some $c > 0$ outside a compact subset of M. Then $\dim H_{(2)}^{n,q}(M, E) < \infty$ for all $q > 0$.*

Proof. Let K_0 be a compact subset of M such that the curvature form Θ_h satisfies $i\Theta_h - c\mathrm{Id}_E \otimes \omega \geq 0$ on $M \setminus K_0$ for some $c > 0$. Then one can find a compact set $K \supset K_0$ and a constant $C > 0$ such that for all $u \in \mathrm{Dom}\bar{\partial} \cap \mathrm{Dom}\bar{\partial}^* \cap L_{(2)}^{n,q}(M, E)$

$$C(\|\bar{\partial}u\|^2 + \|\bar{\partial}^* u\|^2 + \int_K |u|^2 \frac{\omega^n}{n!}) \geq \|u\|^2 \tag{2.40}$$

holds.

Hence, by Lemma 2.9 one can see that the assumption of Proposition 2.2 is satisfied for $H_1 = L_{(2)}^{n,q-1}(M, E)$, $H_2 = L_{(2)}^{n,q}(M, E)$, $H_3 = L_{(2)}^{n,q+1}(M, E)$, $T = \bar{\partial}$ on $L_{(2)}^{n,q-1}(M, E)$, and $S = \bar{\partial}$ on $L_{(2)}^{n,q}(M, E)$. Therefore, by Theorem 2.2, $H_{(2)}^{n,q}(M, E)$ is isomorphic to $\mathrm{Ker}\bar{\partial} \cap \mathrm{Ker}\bar{\partial}^* \cap L_{(2)}^{n,q}(M, E)$ and finite dimensional. $\square$

In virtue of the celebrated unique continuation theorem of Aronszajn [Ar], Theorem 2.36 implies the following.

Corollary 2.10 (cf. [Gra-Ri-1, Gra-Ri-2] and [T-1]). *In the situation of Theorem 2.36 and K_0 as above, assume moreover that $M \neq K_0$, $d\omega = 0$, and that $i\Theta_h \geq 0$ holds everywhere. Then $H_{(2)}^{n,q}(M, E) = 0$ for $q > 0$.*

For the proof of Aronszajn's theorem, see also [B-B-B].

Combining Corollary 2.10 with a theorem of Grauert on the coherence of the direct image sheaves of coherent analytic sheaves by proper holomorphic maps (cf. [Gra-4]), Takegoshi obtained in [T-2] the following:

Theorem 2.37. *Let M be a Kähler manifold and let π be a proper surjective holomorphic map from M to a complex space X. Then, for any Nakano semipositive vector bundle (E, h) over M, the higher direct image sheaves $R^q \pi_* \mathcal{O}(\mathbb{K}_M \otimes E)$ vanish for any $q > n - \dim X$.*

Similarly, we obtain the finiteness counterparts of Theorems 2.10, 2.12 and 2.18 and their strengthened versions as vanishing theorems. However, it is not known to the author whether or not Theorem 2.14 can also be strengthened to a reasonable finite–dimensionality theorem.

2.3.2   Approximation and Isomorphism Theorems

Once one knows the finite-dimensionality of the L^2 cohomology groups, a natural question is to compare them with the ordinary $\bar{\partial}$-cohomology groups, for instance as in the diagram below:

$$
\begin{array}{ccc}
H^{p,q}_{(2)}(M,E) & \dashrightarrow & H^{p,q}_{(2)}(U,E) \\
\downarrow & & \downarrow \\
H^{p,q}(M,E) & \longrightarrow & H^{p,q}(U,E)
\end{array}
\qquad (U\subset\subset M)
$$

Here the map $\dashrightarrow$ is only densely defined.

For the preparation of such a study, let us first go back to the setting of Sect. 2.1, and establish an abstract approximation theorem modelled on a beautiful argument of Hörmander [Hö-1, Proposition 3.4.5] generalizing a well–known proof of Runge's approximation theorem. For that, the following slight extension of the notion of weak convergence is useful.

Definition 2.4. Given a Hilbert space H and a dense subset $V \subset H$, a sequence $u_\mu \in H$ is said to **converge V-weakly** to $u \in H$, denoted by w_V-$\lim_{\mu\to\infty} u_\mu = u$, if $(u, v)_H = \lim_{\mu\to\infty} (u_\mu, v)_H$ holds for any $v \in V$.

Let H_j $(j = 1, 2)$ and $T : H_1 \to H_2$ be as in Sect. 2.1.1. We consider a sequence of such triples (H_1, H_2, T), say $(H_{1,\mu}, H_{2,\mu}, T_\mu)$ $(\mu \in \mathbb{N})$, together with bounded ($\mathbb{C}$ -linear) operators $P_{j,\mu} : H_{j,\mu} \to H_j$ such that the norms of $P_{j,\mu}$ are uniformly bounded and $TP_{1,\mu} = P_{2,\mu}T_\mu$ (in particular $P_{1,\mu}(\mathrm{Dom}T_\mu) \subset \mathrm{Dom}T$) holds for each μ. In this situation, we look for a condition for $\mathrm{Ker}T_\mu$ to approximate $\mathrm{Ker}T$ in some appropriate sense. For that, we fix once for all a dense subset $V \subset H_1$ and require the following:

(i) w_V-$\lim_{\mu\to\infty} P_{1,\mu}P^*_{1,\mu}v = v$ for any $v \in H_1$.
(ii) For any sequence $u_\mu \in \mathrm{Dom}T^*_\mu$ such that w-lim $P_{2,\mu}u_\mu$ and w_V-lim $P_{1,\mu}T^*_\mu u_\mu$ both exist, w_V-lim $P_{1,\mu}T^*_\mu u_\mu = T^*(\text{w-lim } P_{2,\mu}u_\mu)$ holds true.
   Moreover, in accordance with the situation of Sect. 2.1.1, we shall require also:
(iii) There exists a constant $C > 0$ such that $\|u\|_{H_{2,\mu}} \leq C\|T^*_\mu u\|_{H_{1,\mu}}$ holds for all $u \in \mathrm{Dom}T^*_\mu$ $(\mu \in \mathbb{N})$ satisfying $P_{2,\mu}u \perp \mathrm{Ker}T^*$.

Theorem 2.38. *In the above situation, $\sum_\mu P_{1,\mu}(\mathrm{Ker}T_\mu)$ is dense in $\mathrm{Ker}T$.*

Proof. Take any $v \in H_1$ such that $v \perp P_{1,\mu}(\mathrm{Ker}T_\mu)$ for all μ. Then $P^*_{1,\mu}v \perp \mathrm{Ker}T_\mu$ so that, by the assumption (iii) and in virtue of Theorem 2.3.(ii), one can find a constant $\tilde{C} > 0$ and $u_\mu \in \mathrm{Dom}T^*_\mu$ such that $P^*_{1,\mu}v = T^*_\mu u_\mu$ and $\|u_\mu\|_{H_{2,\mu}} \leq \tilde{C}\|v\|_{H_1}$ hold for all μ. Hence, by (i) and (ii), the weak limit, say u, of a subsequence of $P_{2,\mu}u_\mu$ satisfies $T^*u = v$. Hence $v \perp \mathrm{Ker}T$, which proves the assertion. $\square$

Now let (M, ω), (E, h) and K_0 be as in the beginning, and let $U \subset M$ be an open set containing K_0. Given a complete Hermitian metric ω_U on U and a fiber metric h_U of $E|_U$, we shall describe a condition for a sequence ω_μ $(\mu \in \mathbb{N})$ of complete Hermitian metrics on M and a sequence h_μ of fiber metrics of E such that the union of images of $\mathrm{Ker}\bar{\partial} \cap L^{p,q}_{(2)}(M, E, \omega_\mu, h_\mu)$ for all $\mu \in \mathbb{N}$ by the restriction map

$$\rho_U : L^{p,q}_{(2),loc}(M, E) \to L^{p,q}_{(2),loc}(U, E|_U) \tag{2.41}$$

is a dense subset of $\mathrm{Ker}\bar{\partial} \cap L^{p,q}_{(2)}(U, E|_U, \omega_U, h_U)$.

For simplicity, first we assume that $p = n$. Then, $\rho_U(L^{n,q}_{(2)}(M, E, \omega_\mu, h_\mu)) \subset L^{n,q}_{(2)}(U, E|_U, \omega_U, h_U)$ as long as $\omega_\mu \leq \omega_U$ and $h_\mu \geq h_U$ hold on U (cf. the proof of Theorem 2.14). In this setting, a geometric variant of Theorem 2.38 can be stated as follows.

Theorem 2.39. *In the above situation, assume moreover the following:*

(a) ω_μ *are all Kählerian on* $M \setminus K_0$.
(b) $\lim_{\mu \to \infty} \omega_\mu|_U = \omega_U$ *and* $\lim_{\mu \to \infty} h_\mu|_U = h_U$ *locally in the* C^1*-topology.*
(c) *There exists a constant* $c > 0$ *such that* $i\Theta_{h_\mu} - c\mathrm{Id}_E \otimes \omega_\mu \geq 0$ *hold everywhere on* $M \setminus K_0$ *for all* μ.

Then, for all $q \geq 0$,

$$\bigcup_{\mu \in \mathbb{N}} \rho_U(\mathrm{Ker}\bar{\partial} \cap L^{n,q}_{(2)}(M, E, \omega_\mu, h_\mu))$$

is dense in $\mathrm{Ker}\bar{\partial} \cap L^{n,q}_{(2)}(U, E|_U, \omega_U, h_U)$.

Proof. Let $H_1 = L^{n,q}_{(2)}(U, E, \omega_U, h_U)$, $H_2 = L^{n,q+1}_{(2)}(U, E, \omega_U, h_U) \cap \mathrm{Ker}\bar{\partial}$,

$$H_{1,\mu} = L^{n,q}_{(2)}(M, E, \omega_\mu, h_\mu),$$

$$H_{2,\mu} = L^{n,q+1}_{(2)}(M, E, \omega_\mu, h_\mu) \cap \mathrm{Ker}\bar{\partial},$$

$T = \bar{\partial} : H_1 \to H_2$, $T_\mu = \bar{\partial} : H_{1,\mu} \to H_{2,\mu}$, and $P_{j,\mu}$ be the restriction maps. By the assumption that $\omega_\mu \leq \omega_U$ and $h_\mu \geq h_U$, the uniform boundedness of $P_{j,\mu}$ is obvious. It is also clear that (i) and (ii) above hold for $V = C^{n,q}_0(U, E)$ follows from (b). (Note that ω_U is also complete.) To see that (iii) is also true, (a), (b) and (c) are combined as follows.

Suppose that the assertion were false. Then there would exist a nonzero element of $L^{n,q}_{(2)}(U, E, \omega_U, h_U) \cap \mathrm{Ker}\bar\partial$, say v, orthogonal to $\rho_U(\mathrm{Ker}\bar\partial \cap L^{n,q}_{(2)}(M, E, \omega_\mu, h_\mu))$ for all μ. Hence there would exist a sequence $u_\mu \in H_{2,\mu}$ such that $u_\mu|_U \perp \mathrm{Ker}\bar\partial^*$, $\|u_\mu\|_{H_{2,\mu}} = 1$ and $\liminf_{\mu\to\infty} \|T^*_\mu u_\mu\|_{H_{1,\mu}} = 0$, because otherwise v would belong to the image of $\bar\partial^*$. By (b), a subsequence of $u_\mu|_U$ would weakly converge to some element of $L^{n,q}_{(2)}(U, E, \omega_U, h_U) \cap \mathrm{Ker}\bar\partial \cap \mathrm{Ker}\bar\partial^*$, say u. By (a) and (c), $u \neq 0$. But since $u_\mu|_U$ was in the orthogonal complement of $\mathrm{Ker}\bar\partial^*$, so is u. An absurdity!   □

Arguing similarly to the above, from Theorems 2.3 and 2.39 one has the following.

Proposition 2.14. *In the situation of Theorem 2.39, there exists $\mu_0 \in \mathbb{N}$ such that the natural homomorphisms from $H^{n,q}_{(2)}(M, E, \omega_\mu, h_\mu)$ to $H^{n,q}_{(2)}(U, E|_U, \omega_U, h_U)$ induced by ρ_U are injective for all $q > 0$ and $\mu \geq \mu_0$.*

Proof. Let the notation be as in the proof of Theorem 2.39 for $q \geq 0$. Suppose that there exist infinitely many μ such that ρ_U induces noninjective homomorphisms from $H^{n,q+1}_{(2)}(M, E, \omega_\mu, h_\mu)$ to $H^{n,q+1}_{(2)}(U, E|_U, \omega_U, h_U)$. Then, in view of Theorem 2.3, there would exist a sequence $u_\mu \in H_{2,\mu}$ such that $u_\mu|_U \perp (\mathrm{Ker}\bar\partial \cap \mathrm{Ker}\bar\partial^*)$, $\|u_\mu\|_{H_{2,\mu}} = 1$ and $\liminf_{\mu\to\infty} \|T^*_\mu u_\mu\|_{H_{1,\mu}} = 0$, which leads us to a contradiction, similarly to the above.   □

Combining Proposition 2.14 with Theorems 2.39 and 2.38, we obtain:

Theorem 2.40. *In the situation of Theorem 2.39, there exists $\mu_0 \in \mathbb{N}$ such that the natural homomorphisms from $H^{n,q}_{(2)}(M, E, \omega_\mu, h_\mu)$ to $H^{n,q}_{(2)}(U, E|_U, \omega_U, h_U)$ are isomorphisms for all $q > 0$ and $\mu \geq \mu_0$.*

Remark 2.8. A natural question is whether or not the restriction $\mu \geq \mu_0$ is superfluous. It is not, as one can see from the following.

Example 2.9.

$$(M, \omega) = \left(\mathbb{C}, \frac{idz \wedge d\bar z}{(|z|^2 + 1)(\log(|z|^2 + 2))^2}\right),$$

$$E = \bigcup_{z\in\mathbb{C}}\{(\zeta,\xi); z^m\zeta - \xi = 0\}(m \geq 2),$$

$$|(\zeta,\xi)|^2_h = (|\zeta|^2 + |\xi|^2)\frac{\log(|z|^2 + 2)}{|z|^2 + 1},$$

$$(U, \omega_U) = (\{z; |z| < 1\}, i(1 - |z|^2)^{-2}dz \wedge d\bar z)$$

$\Rightarrow$ $H^{1,1}_{(2)}(M, E) \neq 0$ but $H^{1,1}_{(2)}(U, E|U) = 0.$

As this example shows, μ_0 can be arbitrarily large depending on the choices of E, but we do not know any estimate for it in terms of the curvature of E.

Let us briefly illustrate how these approximation and isomorphism theorems are applied.

Proposition 2.15. *Let* $(M, \omega), (E, h), K_0, U, \omega_U$ *and* h_U *be as above, such that* $U = \{x \in M; \phi(x) < d\}$ *for some* C^∞ *plurisubharmonic function* ϕ *on* M. *Then, sequences* ω_μ *and* h_μ *satisfying* $\omega_\mu \leq \omega_U$, $h_\mu \geq h_U$, (a), (b) *and* (c) *exist if* $i\Theta_h - c\mathrm{Id}_E \otimes \omega \geq 0$ *on* $M \setminus K_0$ *holds for some* $c > 0$.

Proof. Put $\omega_U = \omega + i\partial\bar\partial \log \frac{1}{d-\phi}$ and $h_U = h \cdot (d - \phi)$. Let $\lambda_\mu(t)$ ($\mu \in \mathbb{N}$) be a sequence of C^∞ convex increasing functions on $\mathbb{R}$ such that $\lim_{\mu\to\infty} \lambda_\mu(t) = -\log(-t)$ on $(-\infty, 0)$ locally in the C^∞ topology. Then it is easy to see that $\omega_\mu = \omega + i\partial\bar\partial\lambda_\mu(\phi - d)$ and $h_\mu = h \cdot e^{-\lambda_\mu(\phi-d)}$ satisfy the requirements. $\square$

As is easily seen from the above, for pseudoconvex manifolds, the method of detecting the equivalence of L^2 cohomology groups through the L^2 estimates can be naturally extended to establish isomorphism theorems between the ordinary cohomology groups. For instance, let us prove the following.

Theorem 2.41 (cf. [N-R]). *Let* (M, ϕ) *be a pseudoconvex manifold of dimension* n *and let* (E, h) *be a holomorphic Hermitian vector bundle over* M *which is Nakano positive on* $M \setminus M_c$ *for some* c. *Then* $\dim H^{n,q}(M, E) < \infty$ *for all* $q > 0$ *and the restriction homomorphisms* $H^{n,q}(M, E) \to H^{n,q}(M_d, E)$ ($q > 0$) *are isomorphisms for all* $d \geq c$.

Proof. By Theorem 2.40 and Proposition 2.15, the natural restriction homomorphisms

$$\rho_c^d : H^{n,q}_{(2)}(M_d, E) \to H^{n,q}_{(2)}(M_c, E) \quad (q > 0)$$

are isomorphisms if $d > c$. Moreover, since λ_μ in the proof of Proposition 2.15 can be chosen to be of arbitrarily rapid growth, it follows that the natural maps

$$H^{n,q}(M_d, E) \to H^{n,q}_{(2)}(M_c, E)$$

are also bijective.

Injectivity of $H^{n,q}(M, E) \to H^{n,q}(M_c, E)$: Let $u \in L^{n,q}_{(2),loc}(M, E) \cap \mathrm{Ker}\bar\partial$. Suppose that there exists $v \in L^{n,q-1}_{(2),loc}(M_c, E)$ such that $u = \bar\partial v$ holds on M_c. Then, since ρ_c^d are known to be bijective, one can find $v_d \in L^{n,q-1}_{(2),loc}(M_d, E)$ such that $\bar\partial v_d = u$ holds on M_d. By Theorem 2.39, one can then define a sequence $\tilde v_\mu$ ($\mu = 1, 2, \dots$) inductively as follows:

$$\tilde v_1 = v$$

$$\tilde v_2 = v_{c+1} - w_1,$$

where $w_1 \in L_{(2),loc}^{n,q-1}(M_{c+1}, E) \cap \mathrm{Ker}\,\bar\partial$ and $\|w_1 - (v_{c+1} - v_1)\|_{M_c} < \frac{1}{2}$, and

$$v_{\widetilde{\mu+1}} = v_{c+\mu} - w_\mu,$$

where $w_\mu \in L_{(2),loc}^{n,q-1}(M_{c+\mu}, E) \cap \mathrm{Ker}\,\bar\partial$ and $\|w_\mu - (v_{c+\mu} - \widetilde{v_\mu})\|_{M_{c+\mu-1}} < \frac{1}{2^\mu}$. Here the L^2 norm $\|\cdot\|_{M_d}$ on M_d is measured with respect to $\omega + i\partial\bar\partial \log \frac{1}{d-\phi}$ and $h \cdot (d - \phi)$. Then $\bar\partial(\lim \widetilde{v_\mu}) = u$.

Surjectivity of $H^{n,q}(M, E) \to H^{n,q}(M_c, E)$: Let $w \in L_{(2),loc}^{n,q}(M_c, E) \cap \mathrm{Ker}\,\bar\partial$. By Theorem 2.39, one can find $w_\mu \in L_{(2),loc}^{n,q}(M_c + \mu, E) \cap \mathrm{Ker}\,\bar\partial$ similarly to the above, in such a way that $\lim w_\mu$ exists in $L_{(2),loc}^{n,q}(M, E) \cap \mathrm{Ker}\,\bar\partial$ and $\lim (w_\mu|_{M_c}) = w$. $\square$

Since $H^q(M, \mathcal{O}(E)) \cong H^{0,q}(M, \mathbb{K}_M \otimes \mathbb{K}_M^* \otimes E) \cong H^{n,q}(M, \mathbb{K}_M^* \otimes E)$, one has:

Corollary 2.11. *For any strongly pseudoconvex manifold M and for any holomorphic vector bundle E over M, $\dim H^q(M, \mathcal{O}(E)) < \infty$ for any $q > 0$.*

In the situation of Theorem 2.41, it is clear that the above proof shows more precisely that there exists a Hermitian metric ω on M such that $H^{n,q}(M, E) \cong H_{(2)}^{n,q}(M_d, E, \omega + i\partial\bar\partial \log \frac{1}{d-\phi}, h \cdot (d-\phi))$ for all $q > 0$. Similarly, it can be shown also that $H^{n,q}(M, E) \cong H_{(2)}^{n,q}(M_d, E, \omega, h)$ for all $q > 0$ (cf. [Oh-7], where the smoothness assumption on ∂M_d is superfluous). Therefore, one can infer from Corollary 2.11 the following vanishing theorem for ordinary cohomology groups.

Theorem 2.42 (cf. [Gra-Ri-1] and [T-1]). *Let (M, ϕ) and (E, h) be as in Theorem 2.41. Assume moreover that $M \neq M_c$, $d\omega = 0$ and $\Theta_h \geq 0$ on M. Then $H^{n,q}(M, E) = 0$ for $q > 0$.*

Let us also recall a well–known theorem of Grauert which was originally derived from Corollary 2.11. In view of the importance of the result in several complex variables, we shall give a proof as an application of Theorem 2.42.

Theorem 2.43 (cf. [Gra-3]). *Every strongly pseudoconvex manifold is holomorphically convex.*

Proof. Let M and M_c be as in Theorem 2.41, and let $\Gamma = \{x_\mu\}_{\mu=1,2,\dots}$ be any sequence of points in $M \setminus M_c$ which does not have any accumulation point. Let $\pi : \tilde{M} \to M$ be the blow-up of M along Γ and let $\mathscr{I}_\Gamma$ be the ideal sheaf of the divisor $\pi^{-1}(\Gamma)$. Then $\tilde{M}$ is pseudoconvex and it is easy to see that the line bundle $(\mathbb{K}_{\tilde{M}}^* \otimes \mathscr{I}_\Gamma)|_{\pi^{-1}(\Gamma)}$ is positive. Hence

$$H^{n,1}(\tilde{M}, \mathbb{K}_{\tilde{M}}^* \otimes [\pi^{-1}(\Gamma)]^*) \cong H^{n,1}(\pi^{-1}(M_c), \mathbb{K}_{\tilde{M}}^* \otimes [\pi^{-1}(\Gamma)]^*)$$

by Theorem 2.41. Since

$$H^{n,1}(\tilde{M}, \mathbb{K}_{\tilde{M}}^* \otimes [\pi^{-1}(\Gamma)]^*) \cong H^{0,1}(\tilde{M}, [\pi^{-1}(\Gamma)]^*) \cong H^1(\tilde{M}, \mathscr{I}_\Gamma),$$

and $M_c \cap \Gamma = \varnothing$, it follows that the natural restriction map $\mathscr{O}(M) \to \mathbb{C}^\Gamma$ is surjective. This implies the assertion. $\square$

We recall also that Corollary 2.11 was first proved in [Gra-3] by a sheaf theoretic method to derive Theorem 2.43 and later generalized to the following finiteness theorem which has already been mentioned in Chap. 1 (cf. Theorem 1.30).

Theorem 2.44 (Andreotti–Grauert [A-G]). *Let X be a q-convex space and let $\mathscr{F}$ be a coherent analytic sheaf over X. Then $H^q(X, \mathscr{F})$ is finite dimensional for all $p \geq q$.*

Although the above results obtained by the L^2 method do not imply Theorem 2.44 in the full generality, it is by such L^2 "representation" results that analytic methods work effectively in the study of cohomological invariants on complex manifolds and spaces. For instance, let us mention an application of Theorem 2.40 which was observed recently.

Theorem 2.45 (cf. [Oh-34]). *Let M be a compact complex manifold and let D be a smooth divisor of M such that [D] is semipositive. Then, for any holomorphic vector bundle $E \to M$ which is Nakano positive on a neighborhood of D, one can find $\mu_0 \in \mathbb{N}$ such that the restriction homomorphism $H^0(M, \mathscr{O}(\mathbb{K}_M \otimes E \otimes [D]^\mu)) \to H^0(D, \mathscr{O}(\mathbb{K}_M \otimes E \otimes [D]^\mu))$ is surjective for any $\mu \geq \mu_0$.*

To find an effective bound for μ_0 seems to be an interesting question. To the author's knowledge, no purely algebraic proof of Theorem 2.45 is known for the projective algebraic case. It might be worthwhile to note that certain L^2 cohomology is isomorphic to the ordinary cohomology on pseudoconvex manifolds.

Theorem 2.46. *Let (M, ϕ) be a pseudoconvex manifold of dimension n and let (E, h) be a holomorphic vector bundle over M which is Nakano positive on $M \setminus M_d$ for some d. Suppose that there exists $c > 0$ such that $i(\Theta_h - c\mathrm{Id}_E \otimes \Theta_{\mathrm{deth}}) \geq 0$ on $M \setminus M_d$. Then there exists a Hermitian metric ω on M such that*

$$H^{n,q}(M, E) \cong H^{n,q}_{(2)}(M, E, \omega + i\partial\bar{\partial}\phi^2, h \cdot e^{-\phi^2}) \cong$$

$$H^{n,q}_{(2)}(M_d, E, \omega + i\partial\bar{\partial} \log \frac{1}{d - \phi}, h \cdot (d - \phi)) \cong H^{n,q}(M_d, E)$$

for all $q > 0$.

Proof. Let λ_μ $(\mu = 1, 2, \ldots)$ be a sequence of convex increasing functions as in the proof of Proposition 2.15. We are allowed to choose λ_μ so that $\lambda_\mu(t) = \mu(t + \frac{1}{\mu}) + \log \mu$ holds for $t \geq -\frac{1}{\mu}$. Then, for each μ let $\tilde{\lambda}_\mu$ be a C^∞ convex increasing function such that

$$\tilde{\lambda}_\mu(t) = \begin{cases} \lambda_\mu(t) & \text{if } \lambda_\mu(t) \geq t^2 \text{ or } t \leq 0, \\ t^2 & \text{if } \lambda_\mu(t) + 1 < t^2. \end{cases}$$

Then, one can find a Hermitian metric ω on M such that

$$H^{n,q}_{(2)}(M, E, \omega + i\partial\bar{\partial}\phi^2, he^{-\phi^2}) \cong H^{n,q}_{(2)}(M, E, \omega + i\partial\bar{\partial}\widetilde{\lambda_\mu}(\phi - d), he^{-\tilde{\lambda}_\mu(\phi-d)})$$

for any μ.

In fact, one may take $i\Theta_{\det h}$ as ω on $M \setminus M_d$. Since

$$H^{n,q}_{(2)}(M, E, \omega + i\partial\bar{\partial}\tilde{\lambda}_\mu(\phi-d), he^{-\tilde{\lambda}_\mu(\phi-d)}) \cong H^{n,q}_{(2)}(M, E, \omega + i\partial\bar{\partial}\log\frac{1}{d-\phi}, h(d-\phi))$$

for sufficiently large μ, we are done. $\square$

A stronger result holds when E is a line bundle. Namely:

Theorem 2.47. *Let (M, ϕ) be a pseudoconvex manifold of dimension n and let (B, a) be a holomorphic Hermitian line bundle over M which is positive on $M \setminus M_d$ for some $d > 0$. Then there exists a Hermitian metric ω on M such that $H^{p,q}(M, B) \cong H^{p,q}_{(2)}(M, B, \omega + i\partial\bar{\partial}\phi^2, ae^{-\phi^2}) \cong H^{p,q}_{(2)}(M_d, B, \omega + i\partial\bar{\partial}\log\frac{1}{d-\phi}, a(d-\phi))$ if $p + q > n$.*

Sketch of proof. By assumption, there exists a Hermitian metric ω on M such that $\omega = i\Theta_a$ holds on $M \setminus M_d$. The rest is similar to the above. (For the detail, see [Oh-4].) $\square$

If (M, ϕ) is strongly pseudoconvex, then Theorem 2.47 can be strengthened as follows.

Theorem 2.48. *Let (M, ϕ) be a strongly pseudoconvex manifold of dimension n and let (E, h) be a Hermitian holomorphic vector bundle over M. Then there exists a Hermitian metric ω on M such that $H^{p,q}(M, E) \cong H^{p,q}_{(2)}(M_d, E, \omega + i\partial\bar{\partial}\log\frac{1}{d-\phi}, h(d-\phi))$ if $p + q > n$. Here d is any number such that ϕ is strictly plurisubharmonic on $M \setminus M_d$.*

Taking the advantage of strict plurisubharmonicity of ϕ on ∂M_d, a similar argument can be applied to show that $H^{p,q}(M, E) \cong H^{p,q}_{(2)}(M_d, E, \omega + i\partial\bar{\partial}\frac{1}{\log((d-\phi)/R)}, h)$ $(R \gg 1)$ holds for $p + q > n$ (an exercise!).

Combining the isomorphism between the L^2 cohomology and ordinary cohomology with a classical theory of L^2 harmonic forms (cf. [W-2] or [W]), we obtain the following.

Theorem 2.49. *Let (M, ϕ, ω) be a pseudoconvex Kähler manifold of dimension n. If ϕ is strictly plurisubharmonic on $M \setminus M_c$, then $H^r(M, \mathbb{C}) \cong \oplus_{r=p+q}H^{p,q}(M)$ holds for $r > n$ and $H^{p,q}(M) \cong \overline{H^{q,p}(M)}$ for $p + q > n$. Moreover, the map $\wedge^k\omega$: $H^{p-k,q-k}(M) \to H^{p,q}(M)$ defined by $u \mapsto \wedge^k\omega \wedge u$ induces an isomorphism between $H^{p-k,q-k}_0(M)$ and $H^{p,q}(M)$ for $p + q \geq n + 1$ and $k = p + q - n$.*

Corollary 2.12. *Let X be a complex space of dimension n which is nonsingular possibly except at $x \in X$, and let $\tilde{X}$ be a complex manifold which admits a Kähler*

metric and a proper surjective holomorphic map $\pi : \tilde{X} \to X$ *such that* $\pi|_{\tilde{X}\setminus\pi^{-1}(x)}$
is a biholomorphic map. Then there exists a neighborhood $U \ni x$ *such that the r-th
Betti number of* $\pi^{-1}(U)$ *is even for* $r > n$.

Remark 2.9. In the assumption of Corollary 2.12, that $\tilde{X}$ admits a Kähler metric can
be omitted, because there exist a Kähler manifold $\hat{X}$ and a proper bimeromorphic
map $\hat{\pi}; \hat{X} \to \tilde{X}$ obtained by a succession of blow–ups along nonsingular centers in
virtue of Hironaka's fundamental theory of desingularization (cf. [Hn]).

Pursuing an extension of the Hodge theory of this type on strongly pseudoconvex
domains, the following was observed in [Oh-6].

Proposition 2.16 ([Oh-6, Corollary 7 and Note added in proof]). *In the situation
of Corollary 2.12, there exists an arbitrarily small neighborhood* V *of* $\pi^{-1}(x)$ *such
that the restriction homomorphisms* $H^r(V,\mathbb{C}) \to H^r(\partial V,\mathbb{C})$ *are surjective for all*
$r \geq n-1$.

For the proof, the reader is referred to [Oh-5, Oh-6] and [Oh-13]. (See also
[Dm-5], [Sai] and [Oh-T-2].)

∂V is called the **link** of the pair (X,x) if $\partial V = \rho^{-1}(1)$ for some C^∞ function
$\rho : U \to [0,\infty)$ with $V \subset\subset U$ and $(d\rho)^{-1}(0) \cap U = \pi^{-1}(x)$.

Corollary 2.13. $(S^1)^{2n-1}$ *is not homeomorphic to any link if* $n > 1$.

Remark 2.10. In [Ka], it was asked that those 3-manifolds be determined which can
be realized as links of isolated hypersurface singularities in $\mathbb{C}^3$. According to [Ka],
Sullivan has shown that $(S^1)^3$ is not so. A celebrated theorem of Mumford [Mm]
says that $S^3 \not\cong \partial V$ if X is normal and singular at x.

Although the complete metric $\omega + i\partial\bar{\partial}\frac{1}{\log((d-\phi)/R)}$ was useful, the metric $\omega +
i\partial\bar{\partial}\log\frac{1}{d-\phi}$ is also naturally attached to (X,x) (cf. Chap. 4). With respect to this
metric, by extending the Donnelly–Fefferman vanishing theorem (Theorem 2.18),
one has the following in a way similar to that.

Theorem 2.50. *Let* (M,ϕ) *be a strongly pseudoconvex manifold of dimension* n
and let (E,h) *be a Hermitian holomorphic vector bundle over* M. *Then there exists
a Hermitian metric* ω *on* M *such that* $H^{p,q}(M,E) \cong H^{p,q}_{(2)}(M_d,E,\omega + i\partial\bar{\partial}\log\frac{1}{d-\phi},h)$
if $p+q > n$ *and* $H^{p,q}_0(M,E) \cong H^{p,q}_{(2)}(M_d,E,\omega + i\partial\bar{\partial}\log\frac{1}{d-\phi},h)$ *if* $p+q < n$. *Here
d is any number such that ϕ is strictly plurisubharmonic on* $M \setminus M_d$.

Accordingly, the remaining cases $p+q = n$ become of interest. In [D-F], the
following is proved in a slightly more restricted case.

Theorem 2.51. *In the situation of Theorem 2.50,* $\dim H^{p,q}_{(2)}(M_d,E,\omega + i\partial\bar{\partial}\log\frac{1}{d-\phi},h)$
$= \infty$ *if* $p+q = n$.

Proof. See [Oh-12].

Remark 2.11. Note that $H^{p,q}_{(2)}(M_d, E, \omega + i\partial\bar{\partial} \log \frac{1}{d-\phi}, h)$ are Hausdorff. It will be nice if one can show that $\dim H^{p,q}_{(2)}(M) = \infty$ for $p + q = n$ if the metric of M is complete and admits a potential of SBG.

Gromov [Grm] proved:

Theorem 2.52. *Let (M, ω) be a complete Kähler manifold of dimension n. Assume that there exists a C^∞ 1-form τ of bounded length such that $d\tau = \omega$, and that there exists a discrete group Γ of biholomorphic automorphisms of M such that the quotient M/Γ is a compact manifold. Then $\dim H^{p,q}_{(2)}(M, \omega) = \infty$ for $p + q = n$.*

In view of the arguments in the above approximation and isomorphism theorems, it is not so difficult to extend the results to *q-convex* or *q-concave* manifolds. Here we say that a complex manifold M with a C^2 proper map $\psi : M \to (c, 0]$ ($c \in [-\infty, \infty)$) is *q-concave* if ψ is *q-convex* on $\{x; \psi(x) < d\}$ for some $d > c$. In [Hö-1] the following was established by the L^2 method. In fact, it is the prototype of the above arguments.

Theorem 2.53 (Hörmander [Hö-1, Theorem 3.4.9]). *Let (E, h) be a Hermitian holomorphic vector bundle over a complex manifold M of dimension n. If M is q-convex with respect to an exhaustion function ϕ, then $\dim H^q(M, \mathcal{O}(E)) < \infty$. Moreover, if ϕ is q-convex on $M \setminus M_d$ and ∂M_d is smooth, $H^{0,q}(M, E) \cong H^{0,q}_{(2)}(M_d, E, \omega, h)$ holds for any Hermitian metric ω on M. Furthermore, the image of the restriction homomorphism $H^{0,q-1}(M, E) \to H^{0,q-1}_{(2)}(M_d, E, \omega, h)$ is dense. If M is q-concave with respect to an exhaustion function $\psi : M \to (c, 0]$, then $\dim H^{n-q-1}(M, \mathcal{O}(E)) < \infty$ and there exists a Hermitian metric ω on M such that $H^{0,n-q-1}(M, E) \cong H^{0,n-q-1}_{(2)}(M^d, E, \omega, h)$, if ψ is q-convex on $M \setminus M^d$ and ∂M^d is smooth. Here $M^d = \{x; \psi(x) > d\}$.*

Corollary 2.14. *Let M be a q-complete manifold. Then, for any holomorphic vector bundle $E \to M$,*

$$H^{0,p}(M, E) = 0 \quad for \ p \geq q.$$

Combining the techniques originating in [Hö-1] and [A-V-2], which have put the sheaf theoretic development of Oka's solution of the Levi problem by Grauert [Gra-3] and Andreotti and Grauert [A-G] into the framework of the L^2 theory, the following variant of Theorem 2.53 was proved in [Oh-7].

Theorem 2.54. *Let $(E, h), M, \phi$ and ψ be as in Theorem 2.53. If ϕ is q-convex on $M \setminus M_d$, then*

$$H^{0,q}(M, E) \cong H^{0,q}_{(2)}(M_d, E, \omega/(d-\phi)^2, h \cdot e^{-\frac{\alpha}{d-\phi}}) \quad for \ \alpha \gg 1.$$

for any Hermitian metric ω on M. If ψ is q-convex on $M \setminus M^d$, then

$$H^{0,n-q-1}(M,E) \cong H^{0,n-q-1}_{(2)}(M^d, E, \omega + i\frac{\partial\psi \wedge \bar{\partial}\psi}{(d-\psi)^2}, h \cdot e^{-\frac{\alpha}{d-\psi}}) \text{ for } \alpha \gg 1.$$

Since the method of proof is more or less the same as in Theorems 2.41 and 2.46, we shall not repeat it here. The point is that one can choose a Hermitian metric on M in such a way that $(i\partial\bar{\partial}\phi \Lambda u, u) \geq \|u\|^2$ for the (n, q) forms u compactly supported in $M \setminus M_d$ or $(-i\Lambda\partial\bar{\partial}\psi u, u) \geq \|u\|^2$ for the $(0, n-q-1)$ forms u compactly supported in $M \setminus M^d$.

Remark 2.12. In [A-G], the above–mentioned L^2 representation theorem for the $\bar{\partial}$-cohomology is stated as a unique continuation theorem for the sheaf cohomology from sublevel sets (or superlevel sets) of q-convex functions to the whole space. Substantially, the point of the argument is also a Runge–type approximation.

For application of the L^2 approximation technique in (genuine) function theory, the reader is referred to [H-W] and [Sak], for instance.

2.4 Notes on Metrics and Pseudoconvexity

By the methods of L^2 estimates, analytic invariants on complex manifolds have been analyzed above, particularly under the existence of positive line bundles, complete Kähler metrics and plurisubharmonic exhaustion functions. As examples of the situations to which they are applicable, a collection of questions and results in complex geometry related to these basic notions will be reviewed below, mostly without proofs.

2.4.1 Pseudoconvex Manifolds with Positive Line Bundles

We shall review a few results in which pseudoconvex manifolds arise naturally accompanied with positive line bundles.

First, suppose that we are given a closed analytic subset S of a complex space X and a proper surjective holomorphic map π from S to a complex space T. Then a general question is whether or not there exist a complex space Y containing T as a closed analytic subset and a proper surjective holomorphic map from X to Y, say $\tilde{\pi}$ such that $\tilde{\pi}|X \setminus \tilde{\pi}^{-1}(T)$ is a biholomorphic map onto $Y \setminus T$. If it is the case, we shall say that S is **contractible** to T in X. Note that the problem is local along T. Namely, if every point $p \in T$ has a neighborhood U such that $\pi^{-1}(U)$ can be contracted in some neighborhood of it in X, then S is contractible to T in X, by a slight abuse of language. When T is a point, there is a necessary and sufficient condition for the contractibility given by Grauert:

Theorem 2.55 (cf. [Gra-5]). *A compact analytic subset S of X is contractible to a point in X if and only if S admits a strongly pseudoconvex neighborhood system.*

Corollary 2.15. *Let M be a complex manifold of dimension 2 and let $C \subset M$ be a connected analytic subset of dimension one with irreducible components C_j $(1 \leq j \leq m)$. Then C is contractible to a point in M if and only if the matrix $(\deg ([C_j]|_{C_k}))_{1 \leq j,k \leq m}$ is negative definite.*

If X is a complex manifold and S is a compact submanifold, a sufficient (but not necessary) condition for S to have a strongly pseudoconvex neighborhood system is that the normal bundle of S is negative in the sense that its zero section has a strongly pseudoconvex neighborhood system (cf. [Gra-5, Satz 8]). This contractibility criterion is essentially a corollary of Theorem 2.43. Hence a natural question arises whether or not the same is true for the case where S is not compact. When the codimension of S is one, the normal bundle of S is $[S]|_S$, so that its negativity is equivalent to the positivity of $[S]^*$ on a neighborhood U of S which can be chosen to be pseudoconvex by the negativity of $[S]|_S$, by localizing the situation if necessary.

Proposition 2.17. *Let $S \subset X$ be as above. Then, for every point $p \in T$, there exists a pseudoconvex neighborhood U of $\pi^{-1}(p)$ such that $[S]^*|U$ is positive.*

In this way, a pseudoconvex manifold U and a positive line bundle $[S]^*|_U$ arise. Let us mention three results in this situation.

Theorem 2.56 (cf. [N-2] and [F-N]). *Suppose that $\pi : S \to T$ is a complex analytic fiber bundle with fiber $\mathbb{CP}^m$, and that $[S]|_{\pi^{-1}(p)}$ are of degree -1. Then S is contractible to T in X. (Y is actually a manifold.)*

This amounts to a characterization of the blowing–up of Y centered along T. Its proof is done by extending holomorphic functions on S to a neighborhood. In fact, it was for this purpose of contraction that a vanishing theorem like Theorem 2.19 was established. This procedure was generalized as follows.

Theorem 2.57 (cf. [Fj-1]). *Suppose that S is holomorphically convex, $[S]|_S$ is negative, and that $H^1(S, \mathcal{O}(([S]^*)^{\otimes \mu})) = 0$ for all $\mu > 0$. Then S is contractible to T in X.*

In contrast to Theorem 2.43, the condition on the vanishing of the first cohomology groups cannot be omitted. This was clarified by the following.

Theorem 2.58 (cf. [Fj-1, Proposition 3]). *Let B be a positive line bundle over a compact complex manifold F such that $H^1(F, \mathcal{O}(B)) \neq 0$. Then there exists a complex manifold X, a closed submanifold S of X, and a complex analytic fiber bundle $\pi : S \to T$ with fiber F such that S is not contactible to T in X and $[S]|_{\pi^{-1}(p)} \cong B^*$.*

Proof. In the above situation, there exists an affine line bundle $\sigma : \Sigma \to B^*$ such that $\Sigma|_{F_0} \cong F \times \mathbb{C}$ (F_0 = the zero section) but $H^0(U, \mathcal{O}(\Sigma)) = 0$ for any neighborhood $U \supset F_0$. Then the triple ($S = \sigma^{-1}(F_0), X = \sigma^{-1}(U), T = \mathbb{C}$) is such an example. $\square$

Another example of a pseudoconvex manifold with positive line bundles is the quotient of $\mathbb{C}^n$ by the action of a discrete subgroup say Γ satisfying a condition of Riemann type. First let us recall a classical theorem of Lefschetz.

Theorem 2.59 (cf. [B-L], [Kp]). *Assume that $X = \mathbb{C}^n/\Gamma$ is compact and there exists a positive line bundle $L \to X$. Then the following hold.*

(1) $\dim H^0(X, \mathcal{O}(L)) = c_1(L)^n/n!$.
(2) $L^{\otimes 2}$ *is generated by global sections.*
(3) $L^{\otimes 3}$ *is very ample.*

Since the complex semitori $X = \mathbb{C}^n/\Gamma$ are complex Lie groups, they are pseudoconvex (cf. [Mr]). Concerning the positive line bundles on X, they exist if X is compact and the following conditions are satisfied by Γ: there exists a Hermitian form H on $\mathbb{C}^n$ such that:

(i) H is positive definite
    and
(ii) the imaginary part A of H takes integral values on $\Gamma \times \Gamma (\cong H_2(X, \mathbb{Z}))$.

Then, the Appell-Humbert theorem says that, for each **semicharacter** χ,

$$\chi(\gamma + \gamma') = \chi(\gamma)\chi(\gamma')e^{\pi i A(\gamma,\gamma')}, \gamma, \gamma' \in \Gamma \quad by \ definition,$$

one can associate the so-called **factor of automorphy**

$$j(\gamma, z) = \chi(\gamma)e^{\pi H(z,\gamma)+\frac{\pi}{2}H(\gamma,\gamma)} \tag{2.42}$$

and a line bundle $L = L(H, \chi)$ of the form $\mathbb{C}^n \times \mathbb{C}/\Gamma$, where the action of Γ is defined as $\gamma : (z, t) \in \mathbb{C}^n \times \mathbb{C} \to (z + \gamma, j(\gamma, z)t) \in \mathbb{C}^n \times \mathbb{C}$. L is positive because so is H. (For the Appell–Humbert's theorem, see also [B-L] or [Kp]).

In [Ty-2], Theorem 2.59 was extended to the following.

Theorem 2.60. *Let L be a positive line bundle over $X = \mathbb{C}^n/\Gamma$. If X is noncompact, the following hold:*

(1) $\dim H^0(X, \mathcal{O}(L)) = \infty$.
(2) $L^{\otimes 2}$ *is generated by global sections.*
(3) $L^{\otimes 3}$ *is very ample.*

As in the compact case, a semitorus X admits a positive line bundle if its **toroidal reduction** $\hat{X}$ (X is a $\mathbb{C}^a \times (\mathbb{C}^*)^b$-bundle over $\hat{X}$ and $H^0(\hat{X}, \mathcal{O}) = \mathbb{C})$ satisfies a condition similar as above (cf. [A-Gh], [C-C, §2]). It is known that the $\bar{\partial}$-cohomology group of the **toroidal groups** $\hat{X}$ reflect a certain Diophantine property of Γ (cf. [Kz-3] and [Vo]).

For a general pseudoconvex manifold M of dimension $n \geq 2$, positive line bundles are not necessarily ample (cf. [Oh-0]). Nevertheless, it was shown by Takayama [Ty-1] that $K_M \otimes L^m$ is ample if L is a positive line bundle and $m > \frac{1}{2}n(n + 1)$. In the proof of Takayama's theorem, an extension theorem for L^2 holomorphic functions plays an important role (cf. Sect. 3.1.3).

2.4.2   Geometry of the Boundaries of Complete Kähler Domains

In contrast to the vanishing theorems and finiteness theorems on pseudoconvex manifolds, the L^2 vanishing theorems on complete Kähler manifolds were in part motivated by the following theorem of Grauert.

Theorem 2.61 (cf. [Gra-2]). *Let D be a domain in $\mathbb{C}^n$ with real analytic smooth boundary. Then the following are equivalent:*

(1) D admits a complete Kähler metric.
(2) D is pseudoconvex.

That (1) follows from (2) is contained in Proposition 2.12. As for (1) $\Rightarrow$ (2), Grauert showed it by approximating D locally by Reinhardt domains. Real analyticity of ∂D is needed for this argument. In [Oh-2] it was shown under the assumption (2) that, given any point $p \in \mathbb{C}^n \setminus \overline{D}$ and a complex line ℓ intersecting with D and passing through p, there exists a holomorphic function f on $\ell \cap D$ which cannot be continued analytically to p, but extends holomorphically to D by establishing Theorem 2.14 in a special case. To apply this extension argument, C^1-smoothness of ∂D suffices.

Diederich and Pflug [D-P] proceeded further by showing that the purely topological condition $D = \overset{\circ}{\overline{D}}$ (the interior of the closure of D) suffices. For that, they applied Skoda's L^2 division theorem which will be discussed in Chap. 3.

In [Gra-2], it was also shown that the complement of a closed analytic subset of a Stein manifold admits a complete Kähler metric. Indeed, if A is a closed analytic subset of a Stein manifold D, then one can find finitely many holomorphic functions $f_1, \ldots, f_m$ on D such that $A = \{z \in D \; ; \; f_j(z) = 0 \text{ for all } j\}$. Then, for any C^∞ function λ on $D \setminus A$ for which there exists a neighborhood U of A such that $\lambda(z) = -\log\left(-\log\left(\sum |f_j|^2\right)\right)$ holds on $U \setminus A$, there exists a strictly plurisubharmonic exhaustion function ϕ on D such that $i\partial\bar{\partial}(\lambda + \phi)$ is a complete Kähler metric on $D \setminus A$.

In [Oh-3], the following was proved.

Theorem 2.62. *Let D be a pseudoconvex domain in $\mathbb{C}^n$ and let $A \subset D$ be a closed C^1-smooth real submanifold of (real) codimension 2. Then A is a complex submanifold if and only if $D \setminus A$ admits a complete Kähler metric.*

Proof. The "only if" part is already over. Conversely, suppose that $D \setminus A$ admits a complete Kähler metric. To show that A is complex, let $p \in A$ be any point and let ℓ be a complex line intersecting with A transversally at p. Take a Stein neighborhood $W \ni p$ such that $(W \setminus A) \cap \ell$ is biholomorphic to the punctured disc $\{\zeta; 0 < |\zeta| < 1\}$ and $W \setminus A$ is homotopically equivalent to $(W \setminus A) \cap \ell$. Let $\alpha : \widetilde{W \setminus A} \to W \setminus A$ be the double covering. Then, $\widetilde{W \setminus A}$ also admits a complete Kähler metric. By the C^1-smoothness assumption on A, one can apply Theorem 2.14 to extend the single–valued holomorphic function $\sqrt{\zeta}$ on $\alpha^{-1}((W \setminus A) \cap \ell)$ to $\widetilde{W \setminus A}$ with an L^2

growth condition. As a result, one has a holomorphic function on $\widetilde{W \setminus A}$ satisfying an irreducible quadratic equation over $\mathcal{O}(W)$ whose discriminant has zeros or poles along A. Hence A must be complex. □

Theorem 2.62 complements the following well–known result of Hartogs.

Theorem 2.63 (cf. [H]). *Let f be a continuous complex-valued function on a domain $D \subset \mathbb{C}^n$. Then f is holomorphic if and only if the complement of its graph is a domain of holomorphy.*

We note that one can recover its proof by applying Theorem 2.14 to extend a holomorphic function on $\{(z_0, \zeta) \in D \times \mathbb{C}; \zeta \neq f(z_0)\}$ for $z_0 \in D$ with a pole at $\zeta = f(z_0)$ to a holomorphic function on the complement of the graph of f as a meromorphic function on $D \times \mathbb{C}$.

Anyway, from the viewpoint of Oka's solution of the Levi problem, Hartogs's theorem is about the $+\infty$-singular set of plurisubharmonic functions. Conversely, Theorem 2.62 is closely related to the property of the preimages of $-\infty$.

Definition 2.5. A subset F of a complex manifold M is said to be **pluripolar** if there exists a plurisubharmonic function ϕ on M such that $\phi \not\equiv -\infty$ and $F \subset \{z; \phi(z) = -\infty\}$.

Proposition 2.18. *Closed nowhere–dense analytic sets in Stein manifolds are pluripolar.*

Proposition 2.19. *Let D be a domain in $\mathbb{C}^n$ and let $\phi : D \to [-\infty, \infty)$ be a continuous plurisubharmonic function. Then there exists a plurisubharmonic function Φ on D such that Φ is C^∞ on $\Phi^{-1}(\mathbb{R})$ and $\phi^{-1}(-\infty) = \Phi^{-1}(-\infty)$.*

Proof. By a theorem of Richberg, every continuous plurisubharmonic function (with finite values) can be uniformly approximated by C^∞ ones (cf. [R]; see also [Oh-21]). □

Corollary 2.16. *For any pseudoconvex domain $D \subset \mathbb{C}^n$ and a continuous plurisubharmonic function ϕ on D with values in $[-\infty, \infty)$ but not in $\{-\infty\}$, $D \setminus \phi^{-1}(-\infty)$ admits a complete Kähler metric.*

Proof. Take any C^∞ function λ on $D \setminus \phi^{-1}(-\infty)$ satisfying $\lambda(z) = -\log(-\phi(z))$ on $U \setminus \phi^{-1}(-\infty)$. Then $i\partial\bar{\partial}(\Phi + \Psi)$ becomes a complete Kähler metric on $D \setminus \phi^{-1}(-\infty)$ for the above Φ and for some strictly plurisubharmonic exhaustion function Ψ on D. □

Therefore, although under a continuity assumption, Theorem 2.62 gives some information on pluripolar sets.

In this direction, Shcherbina [Shc] has shown a remarkable result:

Theorem 2.64. *Let f be a continuous complex-valued function on a domain $D \subset \mathbb{C}^n$. Then f is holomorphic if and only if its graph is pluripolar.*

The proof is based on a property of *polynomially convex hulls*. (See also [St].)

Coming back to the boundary of complete Kähler domains, a natural question in view of Theorem 2.61 and subsequent remarks is whether or not Theorem 2.62 can be generalized for higher codimensional submanifolds. The answer is yes and no in the following sense.

Theorem 2.65 (cf. [D-F-4, Theorem 1]). *Let A be a closed real analytic subset of a pseudoconvex domain $D \subset \mathbb{C}^n$. Suppose that codimA $\geq$ 3. Then A is complex analytic if and only if $D \setminus A$ admits a complete Kähler metric.*

Theorem 2.66 (cf. [D-F-4, Theorem 2]). *For any integer $k \geq 3$, there exists a closed C^∞ submanifold A of $\{z \in \mathbb{C}^n; \|z\| < 1\}$ such that there exists a complete Kähler metric on the complement of A but A is not complex.*

It was also shown in [D-F-3] that Shcherbina's theorem cannot be generalized to vector–valued functions. Nevertheless, it was shown in [D-F-6] that the submanifold A in Theorem 2.66 still has some distinguished geometric structure. In this series of works, Diederich and Fornaess constructed a smooth real curve in $\mathbb{C}^2$ which is not pluripolar.

As a development from Theorem 2.61, complete Kähler manifolds with curvature conditions have been studied in a wider scope. Some of the results of this type will be reviewed in the next subsection.

2.4.3 Curvature and Pseudoconvexity

If one wants to explore intrinsic properties of noncompact complete Kähler manifolds, it is quite unnatural to presuppose the existence of the boundaries. Namely, we do not see the boundaries of complete manifolds at first. In some cases the boundary appears as a result of compactification (cf. [Sat] and [N-Oh]). Accordingly, in this context concerning the relationship between pseudoconvexity and complete Kähler metric, questions naturally involve the curvature of the metric. It is expected that difference of metric structures implies that of complex structures. A prototype of such a question was solved by Huber [Hu] for Riemann surfaces:

Theorem 2.67. *Let (M, ω) be a noncompact complete Kähler manifold of dimension one whose Gaussian curvature is everywhere positive. Then M is biholomorphically equivalent to $\mathbb{C}$.*

Since any simply connected open (= noncompact) Riemann surface is either $\mathbb{C}$ or $\mathbb{D}(= \{z \in \mathbb{C}; |z| < 1\})$, it is natural to ask for a curvature characterization of the disc $\mathbb{D}$. An answer was given by Milnor in the case where (M, ω) is rotationally symmetric, i.e. when there exists a point $p \in M$ such that with respect to the geodesic length r from p and the associated geodesic polar coordinates r, θ, the Riemann metric associated to ω is of the form $dr^2 + g(r)d\theta^2$. In this case, the Gaussian curvature K is given by $K = -(d^2 g/dr^2)/g$.

Theorem 2.68 (cf. [Ml]). *Let (M, ω) be a simply connected complete Kähler manifold of dimension one. Assume that M is rotationally symmetric. Then the following hold:*

(1) $M \cong \mathbb{C}$ if $K \geq \frac{-1}{r^2 \log r}$ holds for large r.
(2) $M \cong \mathbb{D}$ if $K \leq -\frac{1+\epsilon}{r^2 \log r}$ for large r and $g(r)$ is unbounded.

As for the extension of these results to higher–dimensional cases, Greene and Wu first established the following.

Theorem 2.69 (cf. [G-W-2, Theorem 3]). *Let (M, ω) be a complete noncompact Kähler manifold whose sectional curvature is positive outside a compact set. Then M is strongly pseudoconvex.*

Sketch of proof. Taking the minimal majorant of the *Buseman functions* for all the *rays* emitted from some point, one has a convex exhaustion function on M. For Buseman functions and rays, see [Wu]. □

In [G-W-3], Greene and Wu raised several questions related to the extension of these results. One of them was eventually solved by themselves. The result is very striking:

Theorem 2.70 (cf. [G-W-4], Theorem 4). *Let (M, ω) be a simply connected noncompact complete Kähler manifold of dimension $n \geq 2$. For a fixed $p \in M$, define $k : [0, \infty) \to \mathbb{R}$ by $k(s) = \sup\{|sectional\ curvature\ at\ q|; q \in M, \mathrm{dist}(p, q) = s\}$. Then M is isometrically equivalent to $(\mathbb{C}^n, i \sum dz_j \wedge d\bar{z}_j)$ if the sectional curvature of M is everywhere nonpositive and $\int_0^\infty sk(s)ds < \infty$.*

As for the nonnegatively curved case, they conjectured that a complete Kähler manifold with nonnegative sectional curvature and with positive Ricci curvature is holomorphically convex. Takayama settled it affirmatively in [Ty-3] based on the following.

Theorem 2.71 (cf. [Ty-3, Main Theorem 1.1]). *Pseudoconvex manifolds with negative canonical bundle are holomorphically convex.*

The proof of this beautiful result is actually beyond the scope of the theory presented in Chaps. 1 and 2, and requires a more refined variant of Oka–Cartan theory including construction of specific singular fiber metrics, which will be discussed later in Chap. 3.

As for the Ricci nonpositive case, the following was observed by Mok and Yau [M-Y] in the study of Einstein–Kähler metrics on bounded domains.

Theorem 2.72. *A bounded domain in $\mathbb{C}^n$ is pseudoconvex if it admits a complete Hermitian metric satisfying $-c \leq Ricci\ curvature \leq 0$.*

The proof is based on Yau's version of Schwarz's lemma (cf. [Yau-1]), which is available without the Kählerianity assumption.

2.4.4 Miscellanea on Locally Pseudoconvex Domains

Here we shall collect some of the remarkable facts on locally pseudoconvex domains in or over complex manifolds. Since the proof of the fundamental fact that every locally pseudoconvex Riemann domain over $\mathbb{C}^n$ is an increasing union of strongly pseudoconvex domains depends essentially on the use of the Euclidean metric, it is natural that one needs more differential geometry to analyze locally pseudoconvex domains over complex manifolds.

Let $\pi : D \to \mathbb{CP}^n$ be a locally pseudoconvex noncompact Riemann domain. For any $z \in \mathbb{CP}^n$, let $B(z, r)$ be the geodesic ball of radius r centered at z with respect to the Fubini–Study metric of $\mathbb{CP}^n$, say ω_{FS}. For any $x \in D$ we put

$$\delta(z) = \sup\{r; \pi \text{ maps a neighborhood of } x \text{ bijectively to } B(\pi(x), r)\}.$$

A. Takeuchi extended Oka's lemma (cf. Theorem 1.12) as follows.

Theorem 2.73 (cf. [Tk-1]). $\log \delta(z)^{-1}$ *is plurisubharmonic and* $i\partial\bar{\partial} \log \delta(z)^{-1} \geq \frac{1}{3}\omega_{FS}$ *holds on D.*

Corollary 2.17 (See also [FR]). *Every noncompact locally pseudoconvex domain over $\mathbb{CP}^n$ is a Stein manifold.*

It may be worthwhile to note that the solution of this Levi problem entails the following.

Proposition 2.20. *Let X be a connected compact analytic set of dimension ≥ 1 in $\mathbb{CP}^n$. Then every meromorphic function defined on a neighborhood of X can be extended to $\mathbb{CP}^n$ as a rational function.*

Proof. For any meromorphic function f defined on a domain in $\mathbb{CP}^n$, the maximal Riemann domain to which f is continued meromorphically, i.e. the envelope of meromorphy of f is locally pseudoconvex (see [Siu-2], for instance), so that over $\mathbb{CP}^n$ $(n \geq 2)$ they are either $\mathbb{CP}^n$ or Stein. Since it contains X it must be $\mathbb{CP}^n$. $\square$

Theorem 2.73 was generalized to Riemann domains over Kähler manifolds (cf. [Tk-2], [Suz], [E]).

Definition 2.6. The **holomorphic bisectional curvature** of a Hermitian manifold (M, ω) is a bihermitian form

$$\sum_\mu \left(\sum \Theta^\mu_{\alpha\bar{\beta}\nu} h_{\mu\bar{\kappa}}\right) \xi^\alpha \eta^\nu \overline{\xi^\beta \eta^\kappa} \quad ((\xi^\alpha), (\eta^\nu) \in \mathbb{C}^n \cong T^{1,0}_{M,x}, \quad x \in M)$$

associated to the curvature form $(\Theta^\mu_{\alpha\bar{\beta}\nu})$ of the associated fiber metric $(h_{\mu\bar{\nu}})$ of $T^{1,0}_M$.

Theorem 2.74 (cf. [E] and [Suz]). *Let D be a locally pseudoconvex Riemann domain over a complete Kähler manifold M of positive holomorphic bisectional curvature, and let δ be defined for $D \to M$ similarly to Theorem 2.73. Then $i\partial\bar{\partial} \log \delta^{-1}$ is strictly positive on D.*

Proof (cf. [Oh-18]). Because of the local nature of the problem and in virtue of Oka's lemma, we are allowed to assume that D is a domain with smooth boundary which is everywhere strongly pseudoconvex. Hence, it suffices to consider the situation that ∂D is a complex submanifold of codimension one, the metric on M is real analytic, and $\delta(z)$ is realized by a geodesic say $\gamma : [0, 1] \to M$ joining $z = \gamma(0)$ and a point $\gamma(1) \in \partial D$ in such a way that the length of the geodesic from $\gamma(0)$ to $\gamma(s)$ is s. In this setting, on a neighborhood of $\gamma([0, 1])$ we take a local coordinate $t = (t_1, \ldots, t_n) = (t', t_n)$ such that $s = \mathrm{Re}\ t_n$ on $\gamma([0, 1])$, and look at the Taylor coefficients of the distance from t to $t_n + \sum_{j=1}^{n-1} c_j t_j = 1$, where c_j are so chosen that $\gamma([0, 1])$ is orthogonal to ∂D at $t = (0, \ldots, 0, 1)$. We may assume in advance that c_j are all 0. Then by expressing the Kähler metric, say ω, as

$$\omega = \frac{i}{2} \sum_{j,k=1}^{n} g_{j,\bar{k}} dt_j \wedge d\bar{t}_k \tag{2.43}$$

we have

$$g_{n,\bar{n}}(t) = 1 - \sum_{j,k=1}^{n-1} \lambda_{jk} t_j \bar{t}_k - 4\mathrm{Re}\left(\sum_{j=1}^{n-1} \lambda_{jn} t_j (\mathrm{Im}\ t_n)\right) - 2\lambda_{nn}(\mathrm{Im}\ t_n)^2 + \epsilon(t). \tag{2.44}$$

Here $\lambda_{jk}(= \lambda_{jk}(t', \bar{t}', \mathrm{Re}\ t_n))$ and $\epsilon(t)$ is of order at least 3 in $(t', \bar{t}')$. From (2.38) one can directly read off that

$$\sum_{j,k}^{n} \frac{\partial^2}{\partial t_j \partial \bar{t}_k}\left(\log \frac{1}{\delta(t)}\right)\xi_j \bar{\xi}_j \geq \frac{1}{6}\kappa|\xi|^2 \tag{2.45}$$

holds for any $\xi = (\xi_1, \ldots \xi_n) \in \mathbb{C}^n$, where

$$\kappa = \inf_{t \in \gamma([0,1]), \xi \neq 0} \frac{\left(\sum_{j,k=1}^{n} \lambda_{jk}(t)\xi_j \bar{\xi}_k\right)}{\|\xi\|^2}. \tag{2.46}$$

By the curvature condition on ω (for the complex 2-planes spanned by $\partial/\partial t_n$ and $\partial/\partial t_j$), $\kappa > 0$, from which the desired conclusion is obtained. □

Corollary 2.18. *Every noncompact pseudoconvex Riemann domain over a pseudoconvex Kähler manifold of positive bisectional curvature is Stein.*

Actually this does not generalize Takeuchi's theorem so much, because it turned out that compact Kähler manifolds with positive holomorphic bisectional curvature is biholomorphically equivalent to $\mathbb{CP}^n$ (cf. [M-1], [S-Y]). Nevertheless, as one can see from the above proof, Theorem 2.73 can be immediately extended to the following.

Proposition 2.21. *Every locally pseudoconvex domain in a compact Kähler manifold of semipositive holomorphic bisectional curvature admits a continuous plurisubharmonic exhaustion function.*

Therefore an extension of Corollary 2.17 in this direction is naturally expected. Let us mention two typical results:

Theorem 2.75 (cf. Hirschowitz [Hr]). *Let X be a compact complex manifold whose tangent bundle is generated by global sections. Then every locally pseudoconvex domain in X admits a continuous plurisubharmonic exhaustion function.*

Theorem 2.76 (cf. Ueda [U-1]). *Every noncompact locally pseudoconvex Riemann domain over a complex Grassmannian manifold is Stein.*

Although Hirshcowitz proved more than Theorem 2.75, it is not known whether or not it can be generalized for an arbitary (infinitely sheeted) locally pseudoconvex Riemann domain. A related question of Shafarevitch [Sha] asks whether or not the universal covering spaces of projective algebraic manifolds (or more generally those of compact Kähler manifolds) are holomorphically convex. In the proof of Theorem 2.76, Ueda reduces the question to Oka's theorem for the domains over $\mathbb{C}^n$ by exploiting a result of Matsushima and Morimoto [M-M] which was observed in the study of a question asked by J.-P. Serre. His question was as follows.

By generalizing locally pseudoconvex Riemann domains over complex manifolds or complex spaces, one may consider a complex manifold M paired with a holomorphic map to some complex manifold N, say $f : M \to N$ such that (M, f, N) is locally pseudoconvex in the sense that one can find an open covering U_j of N such that $f^{-1}(U_j)$ are all pseudoconvex. Then it is natural to ask whether or not M is also pseudoconvex (under some reasonable conditions). Within this general setting, the most closely studied case is when N is a Stein manifold and $M \to N$ is a holomorphic (= complex analytic) fiber bundle with Stein fibers. J.-P. Serre asked if M is also Stein. Concerning Serre's problem, several counterexamples (cf. [Sk-3], [Dm-1], [C-L]) and useful partial answers are known.

Theorem 2.77 (cf. [Dm-1]). *There exists a holomorphic $\mathbb{C}^2$ bundle over the unit disc $\mathbb{D}$ which is not Stein.*

Skoda [Sk-5] raises a conjecture that $\mathbb{C}^2$ bundles over $\mathbb{D}$ with polynomial transition functions are Stein.

One of the notable affirmative results is due to N. Mok:

Theorem 2.78 (cf. [Mk]). *Holomorphic fiber bundles over a Stein manifolds with one–dimensional Stein fibers are Stein.*

The reader might notice that, as a variant of Serre's problem we may ask whether or not holomorphic fiber bundles over compact complex manifolds are pseudoconvex. However, there is an immediate counterexample: (the total space of) the line bundle $\mathscr{O}(1)$ over $\mathbb{CP}^n$ is 1-concave! Nevertheless, under some natural geometric circumstances, pseudoconvexity still holds true.

Theorem 2.79 (cf. [D-Oh-2]). *Every holomorphic $\mathbb{D}$-bundle over a compact Kähler manifold is pseudoconvex.*

In this assertion, the fibers can be replaced by any symmetric bounded domain. By such a generalization, a link can be made with the Shafarevitch conjecture (cf. [E-K-P-R]). On the other hand, particularly interesting objects are $\mathbb{D}$-bundles over compact Riemann surfaces since we have:

Theorem 2.80. *A holomorphic $\mathbb{D}$-bundle over a compact Riemann surface is Stein if and only if it has no holomorphic section.*

In [Oh-32], Theorem 2.80 is applied to prove that certain covering spaces over a family of compact Riemann surfaces are holmorphically convex.

Over higher–dimensional compact Kähler manifolds the $\mathbb{D}$-bundles are never Stein as we shall show in Chap. 5 by using the L^2 method. Hence the situation of Theorem 2.80 is really exceptional.

The Kähler condition cannot be dropped in Theorem 2.79 because of the following example.

Example 2.10. (cf. [D-F-5]) Let $\Omega_n = \mathbb{H} \times (\mathbb{C}^n \setminus \{0\})/\Gamma_n$ $(n \geq 2)$, where Γ_n is generated by

$$(\zeta, z_1, \ldots, z_n) \rightarrow (2\zeta, 2z_1, \ldots, 2z_n).$$

Then Ω_n is a $\mathbb{D}$-bundle over a Hopf manifold. Since

$$\Omega_n \cong \left\{ \exp\left(-2\pi^2/\log 2\right) < |\zeta| < 1 \right\} \times (\mathbb{C}^n \setminus \{0\}),$$

Ω_n is not a domain of holomorphy in $\mathbb{C}^n$, so that it does not admit any plurisubharmonic exhaustion function.

Nevertheless, some Hopf manifolds contain open dense Stein subsets, which will also be described in Chap. 5.

Chapter 3
L^2 Oka–Cartan Theory

Abstract Oka–Cartan theory is mainly concerned with the ideals of holomorphic functions on pseudoconvex domains over $\mathbb{C}^n$. To describe how one can find global generators of the ideals, the application of extension theorems and division theorems is indispensable. From the viewpoint of the $\bar{\partial}$-equations, these questions amount to solving those of very special type. Making use of the specific forms of these $\bar{\partial}$-equations, they are solved with precise L^2 norm estimates, yielding optimal quantitative variants of Oka–Cartan theorems.

3.1 L^2 Extension Theorems

From a general point of view, existence theorems and uniqueness theorems for the extension of holomorphic functions are equivalent to the vanishing of cohomology groups with certain boundary conditions, which has already been discussed in Chap. 2. When one wants to study more specific questions of extending functions with growth conditions, such a connection is lost in the sense that the vanishing of cohomology with growth conditions does not imply the existence of extension with growth conditions, except for very special situations. It turns out that there exists a refined L^2 estimate for the $\bar{\partial}$ operator which implies an extension theorem with a right L^2 condition. A general L^2 extension theorem of this kind is formulated on "quasi-Stein" manifolds. They have significant applications in complex geometry.

3.1.1 Extension by the Twisted Nakano Identity

Let (M, ω) be a Kähler manifold of dimension n and let (E, h) be a holomorphic Hermitian vector bundle over M. First let us recall Nakano's identity:

$$[\bar{\partial}, \bar{\partial}_h^{\star}]_{gr} - [\partial^{\star}, \partial_h]_{gr} = [i\Theta_h, \Lambda]_{gr} \tag{3.1}$$

(cf. Theorem 2.7) and a subsequent formula

$$[\bar{\partial}, \theta^*]_{gr} + [\partial^{\star}, \bar{\theta}]_{gr} = [i\partial\bar{\partial}\theta, \Lambda]_{gr} \tag{3.2}$$

© Springer Japan 2015

T. Ohsawa, *L² Approaches in Several Complex Variables*, Springer Monographs in Mathematics, DOI 10.1007/978-4-431-55747-0_3

for any $\theta \in C^{0,1}(M)$ (cf. Theorem 2.8).

To derive a variant of (3.1), we write it as

$$\bar{\partial} \circ \bar{\partial}_h^* + \bar{\partial}_h^* \circ \bar{\partial} - \partial^* \circ \partial_h - \partial_h \circ \partial^* = i(\Theta_h \Lambda - \Lambda \Theta_h). \qquad (3.3)$$

Let η be any positive C^∞ function on M. Then, as a modification of (3.3), one has

$$\bar{\partial} \circ \eta \circ \bar{\partial}_h^* + \bar{\partial}_h^* \circ \eta \circ \bar{\partial} - \partial^* \circ \eta \circ \partial_h - \partial_h \circ \eta \circ \partial^*$$

$$= \bar{\partial}\eta \circ \bar{\partial}_h^* - (\bar{\partial}\eta)^* \bar{\partial} + (\partial\eta)^* \circ \partial_h - \partial\eta \circ \partial^* + i\eta(\Theta_h \Lambda - \Lambda \Theta_h).$$

Hence, applying (3.2) for $\theta = \bar{\partial}\eta$, we obtain

$$\bar{\partial} \circ \eta \circ \bar{\partial}_h^* + \bar{\partial}_h^* \circ \eta \circ \bar{\partial} - \partial^* \circ \eta \circ \partial_h - \partial_h \circ \eta \circ \partial^*$$

$$= \bar{\partial}\eta \circ \bar{\partial}_h^* + \bar{\partial}(\bar{\partial}\eta)^* + (\partial\eta)^* \circ \partial_h + \partial^* \circ \partial\eta - i(\partial\bar{\partial}\eta\Lambda - \Lambda\partial\bar{\partial}\eta - \eta(\Theta_h\Lambda - \Lambda\Theta_h)).$$

Therefore, for any $u \in C_0^{n,q}(M, E)$,

$$\|\sqrt{\eta}\bar{\partial}u\|^2 + \|\sqrt{\eta}\bar{\partial}_h^* u\|^2 \geq 2\mathrm{Re}(\bar{\partial}_h^* u, (\bar{\partial}\eta)^* u) + (i(-\partial\bar{\partial}\eta + \eta\Theta_h)\Lambda u, u). \qquad (3.4)$$

Hence, by the Cauchy–Schwarz inequality one has

$$\|\sqrt{\eta}\bar{\partial}u\|^2 + \|\sqrt{\eta + c^{-1}}\bar{\partial}_h^* u\|^2 \geq (i(-\partial\bar{\partial}\eta - c\partial\eta \wedge \bar{\partial}\eta + \eta\Theta_h)\Lambda u, u) \qquad (3.5)$$

for any positive continuous function c on M. Hence, similarly to Theorem 2.14, we infer from (3.5) the following.

Theorem 3.1. *Let (M, ω) be a Kähler manifold of dimension n, let (E, h) be a holomorphic Hermitian vector bundle over M, and let η be a bounded positive C^∞ function on M. Suppose that there exist a complete Kähler metric on M and positive continuous functions c_1 and c_2 on M such that c_1^{-1} is bounded and*

$$(i(\eta\Theta_h - \partial\bar{\partial}\eta - c_1\partial\eta \wedge \bar{\partial}\eta)\Lambda u, u) \geq (ic_2\partial\eta \wedge \bar{\partial}\eta\Lambda u, u)$$

holds for any $u \in C_0^{n,q}(M, E)$. Then, for any $v \in \mathrm{Ker}\bar{\partial} \cap L_{(2),loc}^{n,q}(M, E)$ of the form $\bar{\partial}\eta \wedge v_0$ such that

$$((ic_2\partial\eta \wedge \bar{\partial}\eta\Lambda)^{-1}v, v) < \infty,$$

one can find a solution to $\bar{\partial}w = v$ satisfying

$$\|(\eta + c_1^{-1})^{-1/2}w\|^2 \leq ((ic_2\partial\eta \wedge \bar{\partial}\eta\Lambda)^{-1}v, v).$$

We note that the boundedness of η and c_1^{-1} is required so that $C_0^{n,q}(M, E)$ is dense in $\mathrm{Dom}(\sqrt{\eta}\bar{\partial}) \cap \mathrm{Dom}(\bar{\partial} \circ \sqrt{\eta + c_1^{-1}})^*$ with respect to the graph norm. It will turn out below that Theorem 3.1 is all one needs to establish an L^2 extension theorem for holomorphic functions in a general and optimal form. Let us show first how it works to prove the following, whose validity itself should be obvious to everybody who knows the area of discs and a few elementary properties of holomorphic functions. As a matter of fact, the length of the proof is the price of generality.

Theorem 3.2. *There exists a holomorphic function f on $\mathbb{D} = \{z \in \mathbb{C}; |z| < 1\}$ such that $f(0) \neq 0$ and*

$$\int_{\mathbb{D}} |f(z)|^2 dx dy \leq \pi |f(0)|^2.$$

Proof. To show the assertion, it suffices to prove that there exists a holomorphic 1-form $f dz$ on $\mathbb{D}$ such that $f(0) = 1$ and

$$\frac{i}{2} \int_{\mathbb{D}} |f(z)|^2 dz \wedge d\bar{z} \leq \pi.$$

Then the problem to be solved is a set of $\bar{\partial}$ equations

$$\bar{\partial} v_\epsilon = \tilde{f} \bar{\partial} \chi_\epsilon(-\log |z|) \wedge dz \tag{3.6}$$

on $\mathbb{D}$ for $0 < \epsilon < 1$, where $\tilde{f}$ is any holomorphic function on $\mathbb{D}$ satisfying $\tilde{f}(0) = 1$ and

$$\chi_\epsilon(t) = \begin{cases} 0 & \text{if } t < -\log \epsilon, \\ \log t - \log(-\log \epsilon) & \text{if } -\log \epsilon \leq t \leq -e \log \epsilon, \\ 1 & \text{if } t > -e \log \epsilon. \end{cases}$$

If we can find solutions v_ϵ such that v_ϵ are extendible to $\mathbb{D}$ continuously to be zero at 0 and

$$\liminf_{\epsilon \to 0} \int_{\mathbb{D}} i v_\epsilon \wedge \overline{v_\epsilon} \leq 2\pi,$$

we are done because a subsequence of $\tilde{f} \chi_\epsilon(-\log |z|) \wedge dz - v_\epsilon$ will tend to a desired extension $f dz$.

To apply Theorem 3.1, we put $M = \{z; 0 < |z| < 1\}$, $(E, h) = (M \times \mathbb{C}, |z|^{-2})$, and

$$\eta_0 = \begin{cases} -\log |z| & \text{if } \epsilon < |z| < 1, \\ (-\log |z|)(\log(-\log |z|)) - \log |z| + (\log(-\log \epsilon))(\log |z|) & \text{if } \epsilon^e \leq |z| \leq \epsilon, \\ (-e \log \epsilon) \log(-e \log \epsilon) - e \log \epsilon + \log(-\log \epsilon)(-e \log \epsilon) & \text{if } |z| < \epsilon^e, \end{cases}$$

The idea here is to *bend* an affine linear function of $-\log(-\log|z|)$ which approximates $\log|z|$ near $|z| = \epsilon$. Then we put $\eta = \eta_0 + \frac{1}{4}(1 - |z|^2)$. Note that $\eta < |z|^{-2}$,

$$\partial\eta_0 = \begin{cases} \partial\log|z| & if\,\epsilon < |z| < 1, \\ -(\log(-\log|z|) + \log(-\log\epsilon))\partial\log|z| & if\,\epsilon^e < |z| \le \epsilon \\ 0, & if\,|z| < \epsilon^e, \end{cases}$$

$-\eta$ is plurisubharmonic, and

$$-\partial\bar\partial\eta = \begin{cases} |z|^2\partial\log|z| \wedge \bar\partial\log|z| & if\ \epsilon < |z| < 1, \\ \frac{\partial\log|z|\bar\partial\log|z|}{-\log|z|} + |z|^2\partial\log|z|\bar\partial\log|z| & if\ \epsilon^e < |z| \le \epsilon. \end{cases}$$

Therefore one has positive C^∞ functions $c = c(\epsilon)$, $b = b(\epsilon)$ and $d = d(\epsilon)$ on M with $\lim_{\epsilon\to 0} b(\epsilon) = 0$, $\lim_{\epsilon\to 0} d(\epsilon) = 0$ such that c^{-1} is bounded, $\eta + c^{-1} = |z|^{-2} + b$ on $\{|z| > \epsilon^e\}$,

$$i(\eta\Theta_h - \partial\bar\partial\eta - c\partial\eta \wedge \bar\partial\eta) \ge 0 \quad \text{everywhere} \tag{3.7}$$

and

$$i(\eta\Theta_h - \partial\bar\partial\eta - c\partial\eta \wedge \bar\partial\eta) \ge i(1-d)\partial\bar\partial\eta \quad \text{on } \{\epsilon^e < |z| < \epsilon\}. \tag{3.8}$$

Therefore, by applying Theorem 3.1 for $c_1 = c$, it is easy to see that one has solutions v_ϵ to (3.6) on $0 < |z| < 1$ which are extendible holomorphically across $\{0\}$, in such a way that their values at 0 are 0 and that

$$\|(\sqrt{\eta + c^{-1}})^{-1}v_\epsilon\|^2 \le 2\pi(1 + A(\epsilon)). \tag{3.9}$$

Here $A(\epsilon) \to 0$ as $\epsilon \to 0$. For these v_ϵ it is clear that

$$\liminf_{\epsilon\to 0} \int_{\mathbb{D}} iv_\epsilon \wedge \overline{v_\epsilon} \le 2\pi.$$

$\square$

It is easy to see that the above proof also works to prove:

Theorem 3.3 (L^2 extension theorem). *Let D be a pseudoconvex domain in $\mathbb{C}^n$ such that $\sup_{z\in D} |z_n| < 1$, let φ be a plurisubharmonic function on D, and let $D' = \{z \in D;\ z_n = 0\}$. Then, for any holomorphic function f on D' satisfying*

$$\int_{D'} |f|^2 e^{-\varphi} d\lambda < \infty,$$

one can find a holomorphic extension $\tilde{f}$ of f to D satisfying

$$\int_D |\tilde{f}|^2 e^{-\varphi} d\lambda \leq \pi \int_{D'} |f|^2 e^{-\varphi} d\lambda.$$

Proof. Since D is pseudoconvex in $\mathbb{C}^n$, there exist an increasing sequence of relatively compact subdomains D_μ, $\mu = 1, 2, \ldots$ whose union is D, and a decreasing sequence of C^∞ plurisubharmonic functions φ_ν on D converging (pointwise) to φ. Therefore, it suffices to show that, for each μ and ν, one can find a holomorphic extension $\tilde{f}_{\mu,\nu}$ of $f|_{D_\mu}$ to D_μ such that

$$\int_{D_\mu} |\tilde{f}_{\mu,\nu}|^2 e^{-\varphi_\nu} d\lambda \leq \pi \int_{D'} |f|^2 e^{-\varphi} d\lambda.$$

But this can be shown similarly to the proof of Theorem 3.2.   □

Remark 3.1. Theorem 3.3 was obtained in [Oh-T-1] in a weaker form in the sense that the estimate for the extension is not in the optimal form as above. The constant π was 1620π there. The optimal form was first proved by Błocki [Bł-2] to settle a question posed by Suita [Su-1]. Related materials will be discussed in more detail in the next chapter.

3.1.2 L^2 Extension Theorems on Complex Manifolds

We shall formulate L^2 extension theorems in more general settings in such a way that they include some of the interpolation theorems in classical complex analysis in one variable. The proofs are essentially the same as in Theorem 3.3.

Definition 3.1. A complex manifold M with a closed subset A is said to be a **quasi-Stein manifold** if A satisfies the following two conditions:

(1) $M \setminus A$ is a Stein manifold.
(2) Each point $x \in A$ has a fundamental neighborhood system $\mathscr{U}_x$ in M such that, for every $U \in \mathscr{U}_x$, the natural restriction map

$$\mathscr{O}(U) \to \{f \in L^{0,0}_{(2),loc}(U); f|_{U \setminus A} \text{ is holomorphic}\}$$

is surjective.

Example 3.1. A complex manifold M with a nowhere–dense closed analytic subset A is quasi-Stein if $M \setminus A$ is a Stein manifold. In particular, a nonsingular projective algebraic variety with a hyperplane section is quasi-Stein.

The following is obvious.

Proposition 3.1. *Let (M, A) be a connected quasi-Stein manifold of dimension $\geq$ 1. Then A is nowhere dense in M.*

Theorem 3.4. *Let (M, A) be a quasi-Stein manifold of dimension n, let w be a holomorphic function on M, let $H = w^{-1}(0)$, and let $H_0 = w^{-1}(0) \setminus dw^{-1}(0)$. Suppose that H_0 is dense in H. Let ϕ and ψ be plurisubharmonic functions on M such that $\sup(\psi + 2\log|w|) \leq 0$. Then, for any holomorphic $(n-1)$-form f on H_0 satisfying $|\int_{H_0} e^{-\phi} f \wedge \bar{f}| < \infty$, there exists a holomorphic n-form F on M such that $F = dw \wedge f$ holds at any point of H_0 and*

$$\left| \int_M e^{-\phi+\psi} F \wedge \bar{F} \right| \leq 2\pi \left| \int_{H_0} e^{-\phi} f \wedge \bar{f} \right|. \tag{3.10}$$

Proof. By the condition (2) on A, it suffices to show the assertion for the manifold $M \setminus A$. Since $M \setminus A$ is Stein, it is an increasing union of strongly pseudoconvex domains. Therefore, we may assume that ϕ and ψ are smooth and it suffices to find, for each relatively compact Stein domain Ω in $M \setminus A$, a holomorphic n-form F_Ω on Ω such that $F_\Omega = dw \wedge f$ holds at any point of $H_0 \cap \Omega$ and

$$\left| \int_\Omega e^{-\phi+\psi} F_\Omega \wedge \overline{F_\Omega} \right| \leq 2\pi \left| \int_{H_0} e^{-\phi} f \wedge \bar{f} \right|. \tag{3.11}$$

To prove this, one has only to repeat the argument of the proof of Theorem 3.2. Namely, by letting $\psi + \log|w|^2$ play the role of $\log|z|^2$ in defining η and by applying the formula (3.5) for $h = e^{-\phi-\log|w|^2}$, we are done. □

Remark 3.2. Theorem 3.3 was first established for the case $\psi \equiv 0$ in [Oh-T-1], where the estimate for F is not sharp. The motivation of introducing ψ in [Oh-17] was to improve the known estimate for the Bergman kernel from below. (For the detail, see the next chapter.) In [Oh-17], $\phi - \psi$ was assumed to be plurisubharmonic. The present form with an optimal estimate was first established in [G-Z-1]. The above proof is taken from [Oh-33].

Now let us proceed to a more general case. Given a positive measure $d\mu_M$ on M, we shall denote by $A^2(M, E, h, d\mu_M)$ the space of L^2 holomorphic sections of E over M with respect to h and $d\mu_M$. Let S be a closed complex submanifold of M and let $d\mu_S$ be a positive measure on S. $A^2(S, E, h, d\mu_S)$ will stand for the space of L^2 holomorphic sections of E over S with respect to h and $d\mu_S$.

Definition 3.2. $(S, d\mu_S)$ is said to be a **set of interpolation** for $A^2(M, E, h, d\mu_M)$ if there exists a bounded linear operator I from $A^2(S, E, h, d\mu_S)$ to $A^2(M, E, h, d\mu_M)$ such that $I(f)|_S = f$ holds for any f.

Let dV_M be any continuous volume form on M. Then we consider a class of continuous functions Ψ from M to $[-\infty, 0)$ such that:

(1) $\Psi^{-1}(-\infty) \supset S$
and

(2) if S is k-dimensional around a point x, there exists a local coordinate $z = (z_1, z_2, \ldots, z_n)$ on a neighborhood U of x such that $z_{k+1} = \ldots = z_n = 0$ on $S \cap U$ and

$$\sup_{U \setminus S} |\Psi(z) - (n - k) \log \sum_{k+1}^{n} |z_j|^2| < \infty.$$

The set of such functions Ψ will be denoted by $\sharp(S)$. Clearly, the condition (2) does not depend on the choices of local coordinates. For each $\Psi \in \sharp(S)$ we define a positive measure $dV_M[\Psi]$ on S as the minimum element of the partially ordered set of positive measures $d\mu$ satisfying

$$\int_{S_k} f d\mu \geq \limsup_{t \to \infty} \frac{2(n - k)}{\sigma_{2n-2k-1}} \int_M f e^{-\Psi} \chi_{R(\Psi, t)} dV_M$$

for any nonnegative continuous function f with $\operatorname{supp} f \Subset M$, where S_k denotes the k-dimensional component of S, σ_m denotes the volume of the unit sphere in $\mathbb{R}^{m+1}$, and $\chi_{R(\Psi, t)}$ the characteristic function of the set

$$R(\Psi, t) = \{x \in M;\ -t - 1 < \Psi(x) < -t\}.$$

Note that the coefficient $2(n - k)/\sigma_{2n-2k-1}$ is chosen in such a way that $d\lambda_z[\log |z_n|^2] = d\lambda_{z'}$ for $z = (z', z_n)$.

Theorem 3.5 (L^2 **extension theorem on manifolds**). *Let M be a complex manifold with a continuous volume form dV_M, let (E, h) be a Nakano semipositive vector bundle over M, let S be a closed complex submanifold of M, let $\Psi \in \sharp(S) \cap C^\infty(M \setminus S)$ and let $\mathbb{K}_M$ be the canonical bundle of M. Then $(S, dV_M[\Psi])$ is a set of interpolation for $A^2(M, E \otimes \mathbb{K}_M, h \otimes (dV_M)^{-1}, dV_M)$ if the following are satisfied:*

(1) There exists a closed subset A such that $S \cap A$ is nowhere dense in S and (M, A) is quasi-Stein.
(2) There exists a positive number δ such that $he^{-(1+\delta)\Psi}$ is a Nakano semipositive singular fiber metric of E.

If, moreover, Ψ is plurisubharmonic on M, the interpolation operator from $A^2(S, E \otimes \mathbb{K}_M, h \otimes (dV_M)^{-1}, dV_M[\Psi])$ to $A^2(M, E \otimes \mathbb{K}_M, h \otimes (dV_M)^{-1}, dV_M)$ can be chosen so that its norm does not exceed $\sqrt{\pi}$.

Sketch of proof. Similarly to Theorem 3.4 it suffices to prove that, for every relatively compact Stein domain Ω in $M \setminus A$, there exists a bounded linear map $I_\Omega : A^2(S, E \otimes \mathbb{K}_M, h \otimes (dV_M)^{-1}, dV_M[\Psi]) \to A^2(\Omega, E \otimes \mathbb{K}_M, h \otimes (dV_M)^{-1}, dV_M)$ whose norm is bounded by a constant independent of Ω such that $I_\Omega(f)|_{S \cap \Omega} = f|_{S \cap \Omega}$ holds for all $f \in A^2(S, E \otimes \mathbb{K}_M, h \otimes (dV_M)^{-1}, dV_M[\Psi])$. When Ψ is plurisubharmonic, the main difference from the situation of Theorem 3.4 is that $S \cap \Omega$ is not necessarily defined as the zero set of a single holomorphic function. However, the assumption on Φ was made in such a way that, as was indicated above by the equation $d\lambda_z[\log |z_n|^2] = d\lambda_{z'}$, Ψ plays the same role as $\log |w|^2$ in the proof of Theorem 3.4. For the proof of the first part, see [Oh-19]. $\square$

Remark 3.3. Theorem 3.5 with a weaker bound $2^4 \sqrt{\pi}$ of the interpolation operator for the last statement was obtained in [Oh-19]. A proof for the bound $\sqrt{\pi}$ was found in [G-Z-2].

Compared to Theorem 3.4, the advantage of Theorem 3.5 is that the condition on the set S is stated in terms of a measure on S and a function with value $-\infty$ along S, since it is often hard to find a generator of the ideal of holomorphic functions vanishing along S. The prototype of Theorem 3.5 is a theorem on interpolation and sampling in one variable due to K. Seip, which we shall recall below. For the proofs the reader is referred to [Sp-1, Sp-2] and [S-W].

Definition 3.3. A subset $\Gamma \subset \mathbb{C}$ is said to be **uniformly discrete** if

$$\inf \{|z - w|; z, w \in \Gamma, z \neq w\} > 0.$$

The **upper uniform density** of a uniformly discrete set Γ is defined to be

$$\limsup_{r \to \infty} \sup_w \frac{\sharp\{z \in \Gamma; |z - w| < r\}}{\pi r^2}$$

which will be denoted by $D^+(\Gamma)$.

For simplicity we put

$$A_\alpha^2 = A^2(\mathbb{C}, \mathbb{C} \times \mathbb{C}, e^{-\alpha|z|^2}, d\lambda_z).$$

Theorem 3.6. *Let Γ be a uniformly discrete subset of $\mathbb{C}$ and let δ_Γ be the Dirac mass supported on Γ. Then, (Γ, δ_Γ) is a set of interpolation for A_α^2 if and only if $\alpha > \pi D^+(\Gamma)$.*

For the unit disc $\mathbb{D}$ we put

$$A_{\alpha,\mathbb{D}}^2 = A^2(\mathbb{D}, \mathbb{D} \times \mathbb{C}, (1 - |z|^2)^\alpha, d\lambda_z).$$

Definition 3.4. A subset $\Gamma \subset \mathbb{D}$ is said to be **uniformly discrete** if

$$\inf \left\{ \left| \frac{z - w}{1 - z\bar{w}} \right|; z, w \in \Gamma, z \neq w \right\} > 0.$$

Letting $\rho(z, w) = |\frac{z-w}{1-z\bar{w}}|$ we put

$$D_{\mathbb{D}}^+(\Gamma) = \limsup_{r \to 1} \sup_z \frac{\sum_{\xi \in \Gamma, \frac{1}{2} < \rho(z,\xi) < r} \log \rho(z, \xi)}{\log (1 - r)}.$$

Theorem 3.7. *Let Γ be a uniformly discrete subset of $\mathbb{D}$. Then $(\Gamma, (1 - |z|^2)\delta_\Gamma)$ is a set of interpolation for $A^2_{\alpha,\mathbb{D}}$ if and only if $\alpha > 2D^+_{\mathbb{D}}(\Gamma)$.*

Theorems 3.6 and 3.7 are related to Theorem 3.5 through an interpretation of the density concepts in the following manner.

For any discrete set $\Gamma \subset \mathbb{C}$ we set

$$C^+(\Gamma) = \inf\{\alpha; \text{ there exists a } \Psi \in \sharp(\Gamma) \text{ such that } \Psi + \alpha|z|^2 \text{ is subharmonic on } \mathbb{C}\}.$$

For a discrete set $\Gamma \subset \mathbb{D}$ we put

$$C^+_{\mathbb{D}}(\Gamma) = \inf\{\alpha; \text{ there exists a } \Psi \in \sharp(\Gamma) \text{ such that } \Psi - \alpha \log(1 - |z|^2) \text{ is subharmonic on } \mathbb{C}\}.$$

Theorem 3.8. *For any uniformly discrete subset $\Gamma \subset \mathbb{C}$ (resp. $\Gamma \subset \mathbb{D}$) one has $\pi D_+(\Gamma) = C^+(\Gamma)$ (resp. $2D^+_{\mathbb{D}}(\Gamma) = C^+_{\mathbb{D}}(\Gamma)$).*

For the proof, see [Oh-19].

3.1.3 Application to Embeddings

Although it might look unlikely from the above exposition, L^2 extension theorems have applications to algebraic geometry. They are used effectively in the induction on the dimension. Uniformity of the estimate in the weight factor φ is important for that purpose. A typical example of such application is to a question of T. Fujita:

Fujita's Conjecture (cf. [F-2]) Let M be a nonsingular projective variety over $\mathbb{C}$ of dimension n, and let $L \to M$ be an ample line bundle. Then:

(1) $\mathbb{K}_M \otimes L^{\otimes(n+1)}$ is generated by global sections
        and
(2) $\mathbb{K}_M \otimes L^{\otimes(n+2)}$ is very ample.

It is obvious that Fujita's conjecture is true if $M = \mathbb{CP}^n$. A theorem of Lefschetz implies its validity for the case of Abelian varieties. There have been a number of results on part (1) of Fujita's conjecture, including its full verification for $n \leq 4$ (cf. [Rd, Ko, E-L, F-3, Km-3]) and its partial verification for general n which requires m to be greater than a constant of order n^2 (cf. [A-S, H-1, H-2, Hei]). As for the connections of Fujita's conjecture to other questions of algebraic geometry, see [M-2] and [Km-1, Km-2, Km-4] as well as [F-2]. In [A-S], Theorem 3.3 (with a weaker estimate) is used to show a semicontinuity property for multiplier ideal sheaves which naturally yields the following effective partial solution to Fujita's conjecture.

Theorem 3.9 (cf. Angehrn and Siu [A-S]). *Let M and L be as above and let κ be a positive number. Suppose that, for any irreducible subvariety W of dimension $1 \leq$*

$d \le n$ in M, the degree $L^d \cdot W$ of $L|W$ satisfies $(L^d \cdot W)^{\frac{1}{d}} \ge \frac{1}{2}n(n+2r-1)+\kappa$. Then the global sections of $\mathscr{O}_M(\mathbb{K}_M \otimes L)$ separate any set of r distinct points $P_1, \ldots, P_r$ of M.

Corollary 3.1. $\mathbb{K}_M \otimes L^m$ is generated by global sections if $m \ge \frac{1}{2}(n^2 + n + 2)$.

Combining Corollary 3.1 with the following lemma, it is known that $\mathbb{K}_M \otimes L \otimes (\mathbb{K}_M \otimes L^m)^{n+1}$ is very ample for $m \ge \frac{1}{2}(n^2 + n + 2)$.

Lemma 3.1. *Let L be a positive line bundle over a compact complex manifold M of dimension n such that $\mathscr{O}_M(L)$ is generated by global sections. Then, $\mathbb{K}_M \otimes L^{n+1} \otimes A$ is very ample for any positive line bundle A.*

The method of Angehrn and Siu was applied by Takayama [Ty-1] to establish a projective embedding theorem for pseudoconvex manifolds with positive line bundles, which amounts to a partial solution to the following generalization of Fujita's conjecture:

Let M be a pseudoconvex manifold of dimension n and let L be a positive line bundle over M.

Generalized Fujita's Conjecture

(1) $\mathbb{K}_M \otimes L^{\otimes(n+1)}$ is generated by global sections
and
(2) $\mathbb{K}_M \otimes L^{\otimes(n+2)}$ is very ample.

Theorem 3.10 (cf. [Ty-1]). *Let M be an n-dimensional pseudoconvex manifold with a positive line bundle L. Then $\mathbb{K}_M \otimes L^m$ is generated by global sections if $m > \frac{1}{2}n(n+1)$, and $(\mathbb{K}_M \otimes L^m)^\ell$ is very ample if $m > \frac{1}{2}n(n+1)$ and $\ell > n+1$.*

The proofs of Theorems 3.9 and 3.10 follows from Nadel's vanishing theorem once one has appropriate singular fiber metrics. Such singular fiber metrics are constructed from the sections of sufficiently high tensor powers of the line bundle L. To obtain a "sufficiently singular" fiber metric by this method, one needs an induction, and at this step the L^2 extension theorem plays a crucial role.

Let us have a glance at this argument by tracing a lemma on the semicontinuity of multiplier ideal sheaves in Siu's exposition [Siu-6]. In the following, "s is a multivalued section of the fractional bundle $L^{\frac{a}{b}}$" means that s^b is a section of the bundle L^a.

Lemma 3.2. *Let M be a compact complex manifold of complex dimension n and let L be a positive line bundle over M. Let P_0 be a point of M and U' be a local holomorphic curve in M passing through P_0 with P_0 as the only (possible) singularity and U be the open unit disc in $\mathbb{C}$ and $\sigma : U \to U'$ be the normalization of U' so that $\sigma(0) = P_0$. Let β be a positive rational number. Let $s_1, \ldots, s_k$ be multivalued holomorphic sections of $pr_1^*(L^\beta)$ over $M \times U$ (pr_1 denotes the projection to the first factor.). Suppose that for almost all $u \in U \setminus \{0\}$ (in the sense that the statement is true up to a subset of measure zero) the point $(\sigma(u), u)$ belongs to the zero-set of the multiplier ideal sheaf of the singular metric $(\sum_{\nu=1}^k |s_\nu|^2)^{-1}|_{M \times \{u\}}$ of*

$L^\beta = pr_1^*(L^\beta)|_{M \times \{u\}}$ (i.e., the function $(\sum_{\nu=1}^k |s_\nu|^2)^{-1}(\cdot, u)$ is not locally integrable at $\sigma(u)$). Then $(P_0, 0)$ belongs to the zero-set of the multiplier ideal sheaf of the singular metric $(\sum_{\nu=1}^k |s_\nu|^2)^{-1}|_{M \times \{0\}}$ of $L^\beta = pr_1^*(L^\beta)|_{M \times \{0\}}$. (i.e., the function $(\sum_{\nu=1}^k |s_\nu|^2)^{-1}(\cdot, 0)$ is not locally integrable at P_0).

Proof. Assume the contrary. Then for some open neighborhood V of P_0 in M the function $(\sum_{\nu=1}^k |s_\nu|^2)^{-1}(\cdot, 0)$ is integrable on V. We can assume without loss of generality that $pr_1^* L|_{V \times U}$ is holomorphically trivial and V is biholomorphic to a bounded pseudoconvex domain in $\mathbb{C}^n$. We apply Theorem 3.3 to the domain $D = V \times U$ and the hyperplane $H = \mathbb{C}^n \times \{0\}$ for $z_n = 0$. For the plurisubharmonic function we use $\varphi = \log(\sum_{\nu=1}^k |s_\nu|^2)$ and for the function to be extended we use $f \equiv 1$. Let F be the holomorphic function on $V \times U$ such that

$$\int_{V \times U} |F|^2 \left(\sum_{\nu=1}^k |s_\nu|^2 \right)^{-1} < \infty$$

and $F(\cdot, 0) = f$ on V. There exist an open neighborhood V' of P_0 in V and an open neighborhood W of 0 in U such that $|F|$ is bounded from below on $V' \times W$ by some positive number. There is a set E of measure zero in W such that $\int_{V' \times \{u\}} (\sum_{\nu=1}^k |s_\nu|^2)^{-1}$ is finite for $u \in W \setminus E$, contradicting the assumption that $(\sum_{\nu=1}^k |s_\nu|^2)^{-1}(\cdot, u)$ is not locally integrable at $\sigma(u)$ for almost all $u \in U \setminus \{0\}$. □

3.1.4 Application to Analytic Invariants

Let $\pi : M \to \mathbb{D}$ be a proper and smooth family of compact complex manifolds and let $M_t = \pi^{-1}(t)$. It was conjectured by S. Iitaka that the **pluricanonical genus** $p_m(M_t) := \dim H^{0,0}(M_t, \mathbb{K}_{M_t}^m)$ does not depend on t if π is a projective morphism (cf. [Nk]). Siu [Siu-7, Siu-9] proved Iitaka's conjecture by combining the L^2 extension theorem and Skoda's division theorem (see the next section). An alternate proof using only the L^2 extension was given by Păun [P].

3.2 L^2 Division Theorems

Given holomorphic functions $g_1, \ldots, g_m$ on a pseudoconvex domain D over $\mathbb{C}^n$, Oka [O-2] proved that, for any holomorphic function h on D which is locally in the ideal generated by $g_j(1 \le j \le m)$, one can find holomorphic functions $f_j(1 \le j \le m)$ on D such that $h = \sum_{j=1}^m f_j g_j$ holds on D. In Cartan's terminology, this is due to the coherence of the kernel of the sheaf homomorphism

$$g : \mathscr{O}^m \to \mathscr{O}, \quad g(u_1, \ldots, u_m) = \sum_{j=1}^{m} u_j g_j$$

and the vanishing of the first cohomology of D with coefficients in the coherent analytic sheaves (cf. [G-R]). The L^2 method of Hörmander [Hö-1, Hö-2] was applied by Skoda [Sk-2] to obtain an effective quantitative refinement of Oka's theorem. After reviewing Skoda's theory in its generalized form (cf. [Sk-4]), we shall show that an L^2 extension theorem on complex manifolds can be applied to prove an L^2 division theorem. This approach has an advantage that it yields a division theorem with an optimal L^2 estimate in some cases.

3.2.1   A Gauss–Codazzi–Type Formula

Let M be a complex manifold and let (E_j, h_j) $(j = 1, 2)$ be two Hermitian holomorphic vector bundles. By a **morphism** between (E_1, h_1) and (E_2, h_2), we shall mean a holomorphic bundle morphism $\gamma : E_1 \to E_2$ such that $\gamma|_{(\text{Ker } \gamma)^\perp}$ fiberwise preserves the length of vectors. Here $(\text{Ker } \gamma)^\perp$ denotes the orthogonal complement of $\text{Ker } \gamma$. Let

$$0 \to S \to E \to Q \to 0 \tag{3.12}$$

be a short exact sequence of holomorphic vector bundles over M and let h be a fiber metric of E. Then one has fiber metrics of S and Q, say h_S and h^Q respectively, for which the arrows in (3.12) become morphisms of Hermitian holomorphic vector bundles. There is a relation between the curvature forms of h, h_S and h^Q which is similar to the classical Gauss–Codazzi formula. It was found by Griffiths [Gri-1, Gri-3]. The presentation below follows [Gri-1, Gri-3] and [Sk-4].

Let D_E be the Chern connection of (E, h) (cf. Chap. 1). Then D_E is decomposed according to the orthogonal decomposition $E = S \oplus Q$:

$$D_E = \begin{pmatrix} D_S & -B^* \\ B & D_Q \end{pmatrix} \tag{3.13}$$

where D_S and D_Q denote respectively the Chern connections of (S, h_S) and (Q, h^Q), $B = \sum B_\alpha dz^\alpha \in C^{1,0}(M, \text{Hom}(S, Q))$, and $B^* = \sum B_\alpha^* d\overline{z}^\alpha \in C^{0,1}(M, \text{Hom}(Q, S))$, where B_α^* denotes the adjoint of B_α. With respect to a local frame $(s_1, \ldots, s_m)$ of E extending a local frame $(s_1, \ldots, s_\ell)$ of S,

$$B_\alpha = (B_{\alpha\nu}^\mu) = \left(\sum_{\sigma=1}^{\ell} h^{\mu\overline{\sigma}} \frac{\partial h_{\nu\overline{\sigma}}}{\partial z_\alpha} \right)$$

In other words, if s and t are respectively smooth sections of S and Q over an open set of M,

$$D_E(s+t) = D_S s - B^* t + B s + D_Q t.$$

B is called the **second fundamental form** of $S \oplus Q$. Note that the second fundamental forms of $S \oplus Q$ and $(S \otimes L) \oplus (Q \otimes L)$ are equal for any holomorphic Hermitian line bundle L.

Example 3.2. Let $M = \mathbb{C}$, let $E = \mathbb{C} \times \mathbb{C}^2$, let $Q = \mathbb{C} \times \mathbb{C}$, let $g : E \to Q$ be given by $(z, (\zeta, \xi)) \to (z, (z\zeta, \xi))$, let $h = (\delta_{\mu\bar{\nu}})$, and let $S = \text{Ker } g$. Then

$$S \oplus Q = \mathbb{C} \cdot (1, -z) \oplus \mathbb{C} \cdot (\bar{z}, 1),$$

$$|1|^2_{h_S} = 1 + |z|^2,$$

$$|1|^2_{h_Q} = (|z|^2 + 1)^{-1},$$

$$B(1, -z) = (0, -dz)$$

and

$$B^*(1) = d\bar{z}.$$

Recall that the curvature form Θ_h is defined as a $\text{Hom}(E, E)$-valued $(1,1)$-form on M satisfying

$$D_E^2 s = \Theta_h s \tag{3.14}$$

for any local C^∞ section s of E. Therefore, from (3.13) one immediately obtains

$$\Theta_h = \begin{pmatrix} D_S^2 - B^* \wedge B & -D(B^*) \\ D(B) & D_Q^2 - B \wedge B^* \end{pmatrix} \tag{3.15}$$

where $D(B)$ denotes the derivative of B with respect to the connection of $\text{Hom}(S, Q)$ associated to D_S and D_Q, namely

$$D_Q(rs) = (Dr)s + r(D_S s)$$

for local sections r of $\text{Hom}(S, Q)$ and s of S. In particular, one has $\bar{\partial}B^* = 0$. From (3.15) we obtain the vector bundle version of the Gauss-Codazzi formula:

$$\begin{cases} \Theta_{h_S} = \Theta_h|_S + B^* \wedge B \\ \Theta_{h_Q} = \Theta_h|_Q + B \wedge B^* \end{cases} \tag{3.16}$$

Here the restrictions of Θ_h are with respect to the orthogonal decomposition $E = S \oplus Q$.

In terms of the local coordinates such that $h = (h_{\mu\bar{\nu}})$ and $B = (\sum_\alpha B^\kappa_{\alpha\nu} dz^\alpha)$,

$$\begin{cases} (\Theta_{hs})^\mu_{\nu\alpha\bar{\beta}} = (\Theta_h)^\mu_{\nu\alpha\bar{\beta}} - \sum_{\kappa,\sigma,\rho} B^\kappa_{\alpha\nu} h^{\mu\bar{\sigma}} \overline{B^\rho_{\beta\sigma}} h_{\kappa\bar{\rho}} \\ (\Theta_{h\varrho})^\kappa_{\rho\alpha\bar{\beta}} = (\Theta_h)^\kappa_{\rho\alpha\bar{\beta}} + \sum_{\mu,\nu,\sigma} B^\kappa_{\alpha\nu} h^{\nu\bar{\mu}} \overline{B^\sigma_{\beta\mu}} h_{\rho\bar{\sigma}} \end{cases} \tag{3.17}$$

Note that the cohomology class in $H^1(M, \mathrm{Hom}(Q, S))$ represented by B^* vanishes if and only if (3.12) splits as a short exact sequence of holomorphic vector bundles. To see this, let g be the given morphism from E to Q and let j be the section of $\mathrm{Hom}(Q, E)$ which satisfies $g \circ j = Id_Q$ and embeds Q into E isometrically. From (3.13), for any C^∞ section t of Q one has

$$\begin{cases} \bar{\partial}(jt) = -B^* t + \bar{\partial}t \\ \bar{\partial}(jt) = (\bar{\partial}j)t + j\bar{\partial}t \end{cases} \tag{3.18}$$

Hence

$$\bar{\partial}j = -B^*.$$

If there exists $A \in C^\infty(M, \mathrm{Hom}(Q, S))$ satisfying

$$\bar{\partial}A = -B^*,$$

then $j - A$ is a holomorphic bundle morphism from Q to E such that $g \circ (j - A) = id_Q$. Now, given a holomorphic section f of Q, one has $\bar{\partial}(jf) = -B^*f$, so that $\bar{\partial}(B^*f) = 0$. If there exists a solution $u \in C^\infty(M, S)$ to the equation

$$\bar{\partial}u = -B^*f, \tag{3.19}$$

$f - u$ will then be holomorphic and satisfy

$$g(f - u) = f. \tag{3.20}$$

Therefore, the problem of lifting holomorphic sections of Q to those of E is reduced to solving the $\bar{\partial}$ equations of the form (3.19) with values in S. In the next subsection, we shall review how Skoda solved a division problem under this formulation.

3.2.2   Skoda's Division Theorem

As well as in the case of extension theorems, it is most appropriate to state a general L^2 division theorem for the bundle-valued $(n, 0)$-forms on n-dimensional complete

Kähler manifolds. Let (M, ω) be a complete Kähler manifold of dimension n and let $0 \to S \to E \to Q \to 0$, $g : E \to Q$, h and B^* be as above. From now on, the ranks of S, E and Q will be denoted by s, p and q, respectively. For the division theorem, Nakano's identity is combined with the following lemma which Skoda called "LEMME FONDAMENTAL".

Lemma 3.3. *Let* $r = \min\{n, s\} = \min\{n, p - q\}$. *For any form* $v \in C^{n,1}(M, S)$ *and* $\beta \in C^{1,0}(M, \mathrm{Hom}(S, Q))$ *one has:*

$$r\langle iTr\beta\beta^* \otimes Id_S \Lambda v, v \rangle \geq |\beta \lrcorner v|^2 \tag{3.21}$$

at every point of M, *where* $Tr\beta\beta^*$ *denotes the trace of* $\beta \wedge \beta^* \in C^{1,1}(M, \mathrm{Hom}(Q, Q))$ *and* $\beta\lrcorner$ *the adjoint of exterior multiplication by* β^*.

Proof. By using the local orthonormal frames, (3.21) follows from the Cauchy–Schwarz inequality

$$\left| \sum_{k=1}^{s} a_k \right|^2 \leq s \sum_{k=1}^{s} |a_k|^2$$

if $r = s$. If $r = n$, (3.21) is reduced to the inequality

$$n \sum_{k,\ell=1}^{s} \left| \sum_{\lambda} \beta_{k\lambda} v_{\ell\lambda} \right|^2 \geq \left| \sum_{k,\lambda} \beta_{k\lambda} v_{k\lambda} \right|^2.$$

For the detail, see [Sk-4, pp. 591–594]. $\qquad\qquad\qquad\qquad\qquad\qquad\square$

In the sense of Nakano positivity similar to the case of curvature form of vector bundles, the lemma says that $irTr\beta\beta^* \otimes Id_S + i\beta\beta^* \geq 0$. Hence, in view of (3.16) and

$$Tr\Theta_{hQ} = \Theta_{\det hQ}$$

one has the following L^2 estimate.

Proposition 3.2. *Assume that* (E, h) *is Nakano semipositive and let* (L, b) *be a Hermitian holomorphic line bundle over* M *whose curvature form* Θ_b *satisfies*

$$i\Theta_b \geq i(r + \epsilon)\Theta_{\det hQ} \tag{3.22}$$

for some $\epsilon > 0$. *Then*

$$\|\bar{\partial} u\|^2 + \|\bar{\partial}^* u\|^2 \geq \frac{\epsilon}{r} \|B \lrcorner u\|^2 \tag{3.23}$$

holds for any $u \in C_0^{n,1}(M, S \otimes L)$.

Example 3.3. In the case of Example 3.2, (3.22) holds for $(L, b) = (\mathbb{C} \times \mathbb{C}, (|z|^2 + 1)^{-1-\epsilon})$.

Hence, solving Eq. (3.19) with an L^2 estimate based on (3.23), Lemma 2.1 and Theorem 2.3, one has:

Theorem 3.11. *Let the situation be as in Proposition 3.2. Then, for any $Q \otimes L$-valued holomorphic n-form f on M which is square integrable with respect to ω and $h^Q \otimes b$, there exists an $E \otimes L$-valued holomorphic n-form e such that $f = g \cdot e$ and*

$$\|e\|^2 \leq \left(1 + \frac{r}{\epsilon}\right) \|f\|^2. \tag{3.24}$$

Corollary 3.2. *Suppose that $(E \otimes \det E, h \otimes \det h)$ is Nakano semipositive and a Hermitian holomorphic line bundle (L, b) over M satisfies*

$$i\Theta_b - i\Theta_{\det h} - i(r + \epsilon)\Theta_{\det h^Q} \geq 0$$

for some $\epsilon > 0$. Then the map

$$H^{n,0}_{(2)}(M, E \otimes L) \rightarrow H^{n,0}_{(2)}(M, Q \otimes L)$$

induced from g is surjective.

Proof. It suffices to apply Theorem 3.11 for the morphism $E \otimes \det E \rightarrow Q \otimes \det E$ and the line bundle $(\det E)^* \otimes L$.   □

Remark 3.4. It was recently shown by Liu, Sun and Yang [L-S-Y] that ample vector bundles have fiber metrics such that $\Theta_{h \otimes \det h}$ is Nakano positive. Satisfying the condition of h in Corollary 3.2. See also [Dm-S].

Applying Corollary 3.2 when M is a bounded pseudoconvex domain in $\mathbb{C}^n$ and $h = (\delta_{\mu\bar{\nu}})$, one has the following.

Corollary 3.3. *Let D be a bounded pseudoconvex domain in $\mathbb{C}^n$, let ϕ be a plurisubharmonic function on D and let $g = (g_1, \ldots, g_p)$ be a vector of holomorphic functions on D. If f is a holomorphic function on D such that*

$$\int_D |f|^2 |g|^{-2k-2-\epsilon} e^{-\phi} d\lambda < \infty$$

holds for $k = \min\{n, p-1\}$ and some $\epsilon > 0$, there exists a vector of holomorphic functions $a = (a_1, \ldots, a_p)$ satisfying

$$f = \sum_{j=1}^{p} a_j g_j$$

and

$$\int_D |a|^2 |g|^{-2k-\epsilon} e^{-\phi} d\lambda < \infty.$$

An advantage of Corollary 3.3 is that it has the following division theorem as an immediate consequence. This special case is useful for the construction of integral kernels (cf. [He]).

Theorem 3.12. *Let D be a bounded domain in $\mathbb{C}^n$ which admits a complete Kähler metric and let $z = (z_1, \ldots, z_n)$ be the coordinate of $\mathbb{C}^n$. Then, for any positive number ϵ, there exists a constant C_ϵ such that, for any holomorphic function f on D satisfying*

$$\int_D |f(z)|^2 |z|^{-2n-\epsilon} d\lambda < \infty,$$

one can find a system of holomorphic functions $a = (a_1, \ldots, a_n)$ satisfying

$$f(z) = \sum_{j=1}^{n} z_j a_j(z)$$

and

$$\int_D |a(z)|^2 |z|^{-2n+2-\epsilon} d\lambda \le C_\epsilon \int_D |f(z)|^2 |z|^{-2n-\epsilon} d\lambda.$$

Corollary 3.4 (cf. [D-P]). *Let D be a domain in $\mathbb{C}^n$ which admits a complete Kähler metric. If $D = \overset{\circ}{\overline{D}}$, then D is a domain of holomorphy.*

Proof. Replacing D by the bounded domains $D \cap \{|z| < R\}$, one may assume that D is bounded in advance. Let z^0 be any point in $\mathbb{C}^n \setminus \overline{D}$. Then $(\sum_{j=1}^{n} |z_j - z_j^0|^2)^{-1}$ is bounded on D, so that by Theorem 3.12 there exist holomorphic functions $a_1(z), \ldots, a_n(z)$ on D satisfying

$$\sum_{j=1}^{n} (z_j - z_j^0) a_j(z) = 1.$$

Hence not all of a_j can be analytically continued to z^0. Since $D = \bar{D}^\circ$, this means that D is a domain of holomorphy. □

Since the method of Skoda is very natural, the estimate in Theorem 3.12 is expected to be optimal. It is indeed the case in some situations as the following example shows.

Example 3.4. The L^2 division problem $zu + v = dz$ on $\mathbb{C}$: It has a solution $(u, v) = (0, dz)$. The squared L^2 norm of this solution $(0, dz)$ with respect to the above-mentioned fiber metric of $E \otimes L$ is $\frac{2\pi}{\epsilon}$, while that of dz with respect to $h^Q b$ is $\frac{2\pi}{1+\epsilon}$. Hence (3.24) is an equality in this case.

However, it is remarkable that Theorem 3.12 is not optimal in the sense that the
following is true.

Theorem 3.13. *Let D be a bounded pseudoconvex domain in $\mathbb{C}^n$. Then there
exists a constant C depending only on the diameter of D such that, for any
plurisubharmonic function ϕ on D and for any holomorphic function f on D
satisfying*

$$\int_D |f(z)|^2 e^{-\phi - 2n \log |z|} d\lambda < \infty,$$

there exists a vector–valued holomorphic function $a = (a_1, \ldots, a_n)$ on D satisfying

$$f(z) = \sum_{j=1}^n z_j a_j(z)$$

and

$$\int_D |a(z)|^2 e^{-\phi(z) - 2(n-1) \log |z|} d\lambda \le C \int_D |f(z)|^2 e^{-\phi(z) - 2n \log |z|} d\lambda.$$

The purpose of the following two subsections is to give a proof of Theorem 3.13
after [Oh-20] as an application of Theorem 3.5.

3.2.3 From Division to Extension

For the proof of Theorem 3.13, we need the following special case of Theorem 3.5.

Theorem 3.14 (Corollary of Theorem 3.5). *Let M, E, S and dV_M be as in
Theorem 3.5. If moreover S is everywhere of codimension one and there exists a
fiber metric b of $[S]^*$ such that $\Theta_h + Id_E \otimes \Theta_b$ and $\Theta_h + (1 + \delta)Id_E \otimes \Theta_b$ are both
Nakano semipositive for some $\delta > 0$, then there exists, for any canonical section s
of $[S]$ and for any relatively compact locally pseudoconvex open subset Ω of M, a
bounded linear operator I from $A^2(S \cap \Omega, E \otimes \mathbb{K}_M, h \otimes (dV_M)^{-1}, dV_M[\log |s|^2])$ to
$A^2(\Omega, E \otimes \mathbb{K}_M, h \otimes (dV_M)^{-1}, dV_M)$ such that $I(f)|_S = f$. Here the norm of I does
not exceed a constant depending only on δ and $\sup_\Omega |s|$.*

Let us describe below how the division problem in Theorem 3.13 is reduced to
an extension problem which can be solved by Theorem 3.14. Let N be a complex
manifold and let F be a holomorphic vector bundle of rank r over N. Let P(F) be
the projectivization of F, i.e. we put

$$P(F) = (F \setminus zero\ section)/(\mathbb{C} \setminus \{0\}).$$

Then $P(F)$ is a holomorphic fiber bundle over N whose typical fiber is isomorphic to $\mathbb{CP}^{r-1}$. Let $L(F)$ be the tautological line bundle over $P(F)$ i.e.

$$L(F) = \coprod_{\ell \in P(F)} \ell$$

where the points of $P(F)$ are identified with complex linear subspaces of dimension one in the fibers of F.

Let $\mathcal{O}(F)$ denote the sheaf of germs of holomorphic sections of F. Then we have a natural isomorphism

$$H^0(N, \mathcal{O}(F)) \cong H^0(P(F^*), \mathcal{O}(L(F^*)^*))$$

which arises from the commutative diagram

$$
\begin{array}{ccc}
L(F^*)^* \longleftarrow \pi^*F \longrightarrow F \\
\searrow \quad \downarrow \quad \downarrow \\
P(F^*) \longrightarrow N
\end{array}
$$

where π denotes the bundle projection (the bottom arrow).

Let $\gamma : F \to G$ be a surjective morphism from F to another holomorphic vector bundle G. Then one has the induced injective holomorphic map

$$P(G^*) \quad \hookrightarrow \quad P(F^*)$$

and a commutative diagram:

$$
\begin{array}{ccc}
L(F^*)^*|_{P(G^*)} & \longleftarrow & \pi^*F|_{P(G^*)} \\
\downarrow \cong & & \downarrow \\
L(G^*)^* & \longleftarrow & \pi^*G|_{P(G^*)}.
\end{array}
$$

One may identify $L(F^*)^*|_{P(G^*)}$ with $L(G^*)^*$ by this isomorphism. Hence, for any holomorphic line bundle L over N, one has a commutative diagram which transfers division problems to extension problems:

$$
\begin{array}{ccc}
H^{0,0}(N, F \otimes L) & \xrightarrow{\;\sim\;} & H^{0,0}(P(F^*), \pi^*L \otimes L(F^*)^*) \\
\downarrow{\scriptstyle \gamma_*} & & \downarrow{\scriptstyle \rho_\gamma} \\
H^{0,0}(N, G \otimes L) & \xrightarrow{\;\sim\;} & H^{0,0}(P(G^*), \pi^*L \otimes L(G^*)^*).
\end{array}
$$

Here ρ_γ denotes the natural restriction map. Note that $P((F \otimes L)^*)$ is naturally identified with $P(F^*)$.

By this diagram, L^2 division problems are also transferred to L^2 extension problems. If γ is a morphism between Hermitian holomorphic vector bundles, one has the following L^2-counterpart of the above:

$$
\begin{array}{ccc}
\mathcal{A}^2(N, F \otimes L) & \xrightarrow{\;\sim\;} & \mathcal{A}^2(P(F^*), \pi^* L \otimes L(F^*)^*) \\
\downarrow{\scriptstyle \gamma_*} & & \downarrow{\scriptstyle \rho_\gamma} \\
\mathcal{A}^2(N, G \otimes L) & \xrightarrow{\;\sim\;} & \mathcal{A}^2(P(G^*), \pi^* L \otimes L(G^*)^*)).
\end{array}
$$

Given a volume form dV on N and a fiber metric h of F, the volume form on $P(F^*)$ associated to dV and h is defined as

$$
dV_h = \bigwedge^{r-1} (i\partial\bar{\partial} \log |\zeta|_h^2) \wedge dV
$$

where ζ denotes the fiber coordinate of F. In order to apply Theorem 3.14 for $M = P(F^*)$ and $S = P(G^*)$, the condition on the codimension is missing in general. To fill this gap, let us replace $P(F^*)$ by its monoidal transform $\sigma : \widetilde{P(F^*)} \to P(F^*)$ along $P(G^*)$ and consider the restriction map

$$
A^2(\widetilde{P(F^*)}, \sigma^* L(F^*)^*) \;\to\; A^2(\sigma^{-1}(P(G^*)), \sigma^* L(G^*)^*)
$$

or equivalently the map

$$
H^{n+r-1,0}_{(2)}(\widetilde{P(F^*)}, \sigma^* L(F^*)^* \otimes \mathbb{K}^*_{\widetilde{P(F^*)}}) \;\to\;
$$

$$
H^{0,0}_{(2)}(\sigma^{-1}(P(G^*)), \sigma^* L(G^*)^* \otimes \mathbb{K}^*_{\widetilde{P(F^*)}} \otimes \mathbb{K}_{\widetilde{P(F^*)}})
$$

Here the volume form on $\widetilde{P(F^*)}$ is induced from dV_h and a fiber metric b of the bundle $[\sigma^{-1}(P(G^*))]$ via the isomorphism

$$
\mathbb{K}_{\widetilde{P(F^*)}} \cong \sigma^* \mathbb{K}_{P(F^*)} \otimes [\sigma^{-1}(P(G^*))]^{\otimes(k-1)}
$$

where k is the codimension of $P(G^*)$ in $P(F^*)$. Accordingly, as the fiber metric of $\sigma^* L(F^*)^* \otimes \mathbb{K}^*_{\widetilde{P(F^*)}}$ we take $\sigma^*(\pi^* h \cdot dV_h) \cdot b^{k-1}$.

3.2.4 Proof of a Precise L^2 Division Theorem

Let the situation be as in the hypothesis of Theorem 3.13. We may assume that ϕ is smooth since D is Stein. Since the assertion is obviously true if $n = 1$ (even for any ϕ), we assume that $n \geq 2$. For simplicity we shall assume that $\phi = 0$, since the proof is similar for the general case.

To apply Theorem 3.14 we put

$$N = D \setminus \{0\},$$

$$dV = \bigwedge^{n} (i\partial\bar{\partial}(|z|^2 + \log|z|^2)),$$

$$F = N \times \mathbb{C}^n,$$

$$h = (\delta_{\mu\bar{\nu}}),$$

$$G = N \times \mathbb{C}$$

and

$$\gamma(z, \zeta) = \left(z, \sum_{j=1}^{n} z_j \zeta_j\right).$$

To find a right fiber metric b of $[\sigma^{-1}(P(G^*))]$ one needs a little more geometry.

First we consider the extensions

$$\hat{\pi} : \hat{F} = \mathbb{C}^n \times \mathbb{C}^n \to \mathbb{C}^n, \qquad \hat{G} = (\mathbb{C}^n \setminus \{0\}) \times \mathbb{C}$$

of the above bundles F and G and note that the closure of the image of $P(\hat{G}^*)$ in $P(\hat{F}^*)$, say $\overline{P(\hat{G}^*)}$, is nothing but the monoidal transform of $\mathbb{C}^n$ with center 0.

Let $\hat{\sigma} : \widetilde{P(\hat{F}^*)} \to P(\hat{F}^*)$ be the monoidal transform along $\overline{P(\hat{G}^*)}$. Observe that, for any complex line ℓ in the projectivization of $\hat{\pi}^{-1}(0)$, the normal bundle $N_{\overline{P(\hat{G}^*)}/P(\hat{F}^*)}$ satisfies

$$\mathcal{O}(N_{\overline{P(\hat{G}^*)}/P(\hat{F}^*)})|\ell \cong \mathcal{O}^{n-2} \oplus \mathcal{O}(1).$$

Therefore $[\hat{\sigma}^{-1}(\overline{P(\hat{G}^*)})]^*$ admits a fiber metric b such that, with respect to the induced fiber metric $\hat{h}$ of $\hat{\sigma}^* L(\hat{F}^*)^*$,

$$i\Theta_b + (1 + \epsilon)i\Theta_{\hat{h}} > 0 \tag{3.25}$$

holds for any $\epsilon > 0$.

On the other hand,

$$\mathbb{K}_{\widetilde{P(\hat{F}^*)}} \cong \hat{\sigma}^* \mathbb{K}_{P(\hat{F}^*)} \otimes [\hat{\sigma}^{-1}(P(\hat{G}^*))]^{\otimes(n-2)},$$

so that

$$\hat{\sigma}^* L(\hat{F}^*)^* \otimes \mathbb{K}^*_{\widetilde{P(\hat{F}^*)}} \cong \hat{\sigma}^* L(\hat{F}^*)^* \otimes \hat{\sigma}^* \mathbb{K}_{P(\hat{F}^*)} \otimes [\hat{\sigma}^{-1}(P(\hat{G}^*))]^{\otimes(n-2)}.$$

Since $i\Theta_{\hat{\sigma}dV_h} \geq ni\Theta_{\hat{h}}$ follows immediately from the definition of dV_h, one has

$$i\Theta_{\hat{h}\otimes\hat{\sigma}^*dV_h\otimes b^{n-2}} + (1+\delta)i\Theta_b \geq (1+n)i\Theta_{\hat{h}} + (n-1+\delta)i\Theta_b$$

on $\widetilde{P(\hat{F}^*)} \setminus \hat{\sigma}^{-1}(\hat{\pi}^{-1}(0))$. By (3.25) the right–hand side of the above inequality is positive if $1 - n < \delta < 2$. Hence Theorem 3.14 is applicable if $n \geq 2$. $\square$

Remark 3.5. From the above proof, the difference of the weights in the L^2 estimate for the solution is geometrically understood as the singularity of the induced fiber metric of $N \times \mathbb{C}$ over 0.

It might be worthwhile to compare Theorem 3.13 with its predecessor obtained by Skoda in [Sk-2]:

Theorem 3.15. *Let D be a pseudoconvex domain in $\mathbb{C}^n$, let ϕ be a plurisubharmonic function on D and let $g = (g_1, \ldots, g_p)$ be a vector of holomorphic functions on D and let f be a holomorphic function on D such that*

$$\int_D |f|^2 |g|^{-2k-2}(1 + \Delta \log |g|)e^{-\phi}d\lambda < \infty$$

holds for $k = \min\{n, p-1\}$, where Δ denotes the Laplacian. Then there exists a vector of holomorphic functions $a = (a_1, \ldots, a_p)$ satisfying

$$f = \sum_{j=1}^{p} a_j g_j$$

and

$$\int_D |a|^2 |g|^{-2k}(1 + |z|^2)^{-2}e^{-\phi}d\lambda < \infty.$$

The author does not know whether or not one can get rid of the factor $1 + \Delta \log |g|$ from the above condition, although it is certainly the case when $g = z$ as Theorem 3.13 shows.

Skoda's L^2 division theory, as well as the L^2 extension theorems inspired by it, was meant to refine the Oka-Cartan theory of ideals of analytic functions. As a result, it has applications to subtle questions in algebra. For instance, Theorem 3.11

can be applied to estimate the degrees of the polynomial solutions $f = (f_1, \ldots, f_p)$ to

$$\sum_{j=1}^{p} f_j g_j = 1$$

for the polynomials g_j without common zeros in $\mathbb{C}^n$ (cf. [B-G-V-Y]). In the next section, we shall give a survey on applications of the L^2 method to the ideals in $\mathbb{C}\{z\}$ which started from the breakthrough in [B-Sk].

3.3 L^2 Approaches to Analytic Ideals

Beginning with a celebrated application of Skoda's division theorem to a refinement of Hilbert's Nullstellensatz, we shall review subsequent results on the ideals in $\mathbb{C}\{z\}$ obtained by the L^2 method, particularly those on the multiplier ideal sheaves in $\mathscr{O}_{\mathbb{C}^n}$. They are initiated by Nadel [Nd] and enriched by Demailly and Kollár [Dm-K] and Demailly, Ein and Lazarsfeld [Dm-E-L]. Recent activity of Berntdsson [Brd-2] and Guan and Zhou [G-Z-2, G-Z-3, G-Z-5] settled a question posed in [Dm-K] (see also Hiep [Hp]).

3.3.1 Briançon–Skoda Theorem

In [B-Sk], Briançon and Skoda extended Euler's identity

$$rf = \sum_{j=1}^{n} z_j \frac{\partial f}{\partial z_j}$$

which holds for any homogeneous polynomial f of degree r, by establishing a remarkable result on the *integral closure* of ideals in $\mathbb{C}\{z\}$. Recall that the **integral closure** $\overline{\mathscr{I}}$ of an ideal $\mathscr{I}$ of a commutative ring $\mathscr{R}$ is defined as

$$\overline{\mathscr{I}} = \{x \in \mathscr{R} \; ; \quad \text{there exists a monic polynomial } b(X) = X^{q+1} + \sum_{j=0}^{q} b_j X^j$$

such that $b_j \in (\mathscr{I})^{q+1-j} \quad (j = 0, \ldots, q)$ and $b(x) = 0\}$. (3.26)

Theorem 3.16 (Briançon–Skoda theorem). *For any ideal $\mathscr{I} \subset \mathbb{C}\{z\}$ which is generated by k elements, $\overline{\mathscr{I}^{k+\ell-1}} \subset \mathscr{I}^\ell$ holds for any $\ell \in \mathbb{N}$. Moreover, $\overline{\mathscr{I}^{n+\ell-1}} \subset \mathscr{I}^\ell$ if $k \geq n$.*

Corollary 3.5. *Let f be any element of $\mathbb{C}\{z\}$ without constant term and let $\mathscr{I}_f$ be the ideal generated by $z_1 \frac{\partial f}{\partial z_1}, \ldots, z_n \frac{\partial f}{\partial z_n}$. Then $f^{\ell+n-1} \in (\mathscr{I}_f)^\ell$ for any nonnegative integer ℓ.*

Corollary 3.5, which we shall prove below, had been conjectured by J. Mather (cf. [Wl]). For the systematic treatment including the proof of Theorem 3.16, the reader is referred to [B-Sk] or [Dm-8]. For non-L^2 proofs, see [L-T, S] and [Sz].

Lemma 3.4. *Let $f, g_1, \cdots, g_k$ be germs of holomorphic functions vanishing at $0 \in \mathbb{C}^n$. Suppose that for every holomorphic map $\gamma : \mathbb{D} \to \mathbb{C}^n$ with $\gamma(0) = 0$ one can find a positive number C_γ such that $|f \circ \gamma| \le C_\gamma |g \circ \gamma|$ holds on a neighborhood of $0 \in \mathbb{D}$. Then there exists a constant C such that $|f| \le C|g|$ holds on a neighborhood of $0 \in \mathbb{C}^n$.*

Proof. Let $(A, 0)$ be the germ of an analytic set in $(\mathbb{C}^{n+k}, 0)$ defined by

$$g_j(z) = f(z)z_{n+j}, \quad 1 \le j \le k.$$

If one could not find C, there would exist a sequence p_μ converging to the origin such that $f(p_\mu) \ne 0$ and $\lim |g(p_\mu)|/|f(p_\mu)| = 0$. Then, taking a germ of a holomorphic map from $(\mathbb{C}, 0)$ to $(\mathbb{C}^{n+r}, 0)$ whose image is contained in $(A, 0)$ but not in $f^{-1}(0)$, one has a holomorphic curve as the projection to the first n factors, which contradicts the assumption.   □

Proof of Corollary 3.5. Let $g = (z_1 \frac{\partial f}{\partial z_1}, \ldots, z_n \frac{\partial f}{\partial z_n})$. By Lemma 3.4 and the chain rule for differentiation, it is easy to see that $|f| \le C|g|$ holds on a neighborhood of 0. Hence the conclusion follows from Corollary 3.2, since

$$\int_U |g|^{-\epsilon} d\lambda < \infty$$

for sufficiently small ϵ for a sufficiently small neighborhood $U \ni 0$ if f is any nonzero element of $\mathbb{C}\{z\}$ without constant term.   □

Remark 3.6. A connection between Corollary 3.5 and a topological theory of isolated hypersurface singularities was suggested by E. Brieskorn and established by J. Scherk [Sch]. The ideal $\mathscr{J}$ generated by $\frac{\partial f}{\partial z_1}, \ldots, \frac{\partial f}{\partial z_n}$ is called the **Jacobian ideal** of f. $\mathscr{J}$ plays an important role in the theory of period mappings (cf. [Gri-1]).

As for Theorem 3.16, it was extended by Demailly [Dm-8] to a result on multiplier ideal sheaves. A weak form of it says that $\mathscr{I}_{a^\ell} \subset \mathscr{I}_a^{\ell-n}$ for any singular Hermitian line bundle (B, a) on a complex manifold of dimension n, and a strengthened version established in [Dm-E-L] says that

$$\mathscr{I}_{a_1 a_2} \subset \mathscr{I}_{a_1} \cdot \mathscr{I}_{a_2}$$

for any (B_1, a_1) and (B_2, a_2) (subadditivity theorem). Since results of this kind have applications to algebraic geometry, there have been subsequent developments in the theory of multiplier ideal sheaves. In the next subsection, we shall review some of them which are related to the L^2 theory.

3.3.2 Nadel's Coherence Theorem

Before describing the results on the multiplier ideal sheaves, let us present the most basic result by Nadel [Nd]. It is the coherence of multiplier ideal sheaves.

Concerning the vanishing theorems for the $\bar{\partial}$–cohomology, global theorems look similar to local theorems from the L^2 viewpoint, since the geometric conditions needed are positivity of the bundle metric and (complete) Kählerianity of the base. On the other hand, in the finite-dimensionality theorems, geometry is involved in a subtler way. (Recall Theorems 2.32 and 2.36, for instance.) Nadel's coherence theorem is a local theorem attached to his vanishing theorem (cf. Theorem 2.24).

Theorem 3.17 (Nadel's coherence theorem [Nd]). *For any singular fiber metric a of a holomorphic line bundle B over a complex manifold M, $\mathscr{I}_a$ is a coherent ideal sheaf of $\mathscr{O}_M$.*

Proof. Since the assertion is local, we may assume that M is a bounded Stein domain in $\mathbb{C}^n$ and $a = e^{-\varphi}$ for some plurisubharmonic function φ. Let $\mathscr{I}$ denote the ideal sheaf generated by the global sections of $\mathscr{I}_a$. Since the ideal sheaves generated by finitely many global sections of $\mathscr{I}_a$ are coherent (Oka's coherence theorem), and since $\mathbb{C}\{z\}$ is a Noetherian ring, $\mathscr{I}$ is coherent. Therefore it remains to show that $\mathscr{I}_{a,x} = \mathscr{I}_x$ holds for any $x \in M$. Since $\mathscr{O}_{M,x}(\cong \mathbb{C}\{z\})$ is Noetherian, by the intersection theorem of Krull it suffices to show that

$$\mathscr{I}_x + \mathscr{I}_{a,x} \cap m_x^k = \mathscr{I}_{a,x} \tag{3.27}$$

for every $k \in \mathbb{N}$ (cf. [Ng, Chapter 1, Theorem 3.11]). But (3.27) is obtained immediately by applying Theorem 2.24.   □

3.3.3 Miscellanea on Multiplier Ideals Sheaves

Since the results on the multiplier ideal sheaves are all local in this subsection, we shall consider only trivial line bundles over complex manifolds and denote the sheaves $\mathscr{I}_{e^{-\varphi}}$ by $\mathscr{I}(\varphi)$ for simplicity.

A striking variant of Briançon–Skoda's theorem is a subadditivity theorem due to Demailly, Ein and Lazarsfeld [Dm-E-L]. It is obtained by combining the following two basic formulae.

Restriction Formula *Let M be a complex manifold, let φ be a plurisubharmonic function on M, and let S be a closed complex submanifold of M. Then*

$$\mathscr{I}(\varphi|_S) \subset \mathscr{I}(\varphi)|_S.$$

Proof. A direct consequence of Theorem 3.3.   □

Addition Formula *Let M_1, M_2 be complex manifolds, $\pi_j : M_1 \times M_2 \to M_j$, $j = 1, 2$ the projections, and let φ_j be a plurisubharmonic function on M_j. Then*

$$\mathscr{I}(\varphi_1 \circ \pi_1 + \varphi_2 \circ \pi_2) = \pi_1^* \mathscr{I}(\varphi_1) \cdot \pi_2^* \mathscr{I}(\varphi_2).$$

Proof. It suffices to show the assertion when M_j are bounded Stein domains in complex number spaces. Consider the ideal sheaf, say $\mathscr{J}$ generated by global sections of $\pi_1^* \mathscr{I}(\varphi_1) \cdot \pi_2^* \mathscr{I}(\varphi_2)$. By Fubini's theorem, it is easy to see that the orthogonal complement of the subspace of the square integrable sections of $\mathscr{J}$ consisting of the square integrable sections of $\pi_1^* \mathscr{I}(\varphi_1) \cdot \pi_2^* \mathscr{I}(\varphi_2)$ is 0. Hence, similarly to the proof of Theorem 3.14, one has the asserted equality in the sheaf level.   □

Theorem 3.18 (Subadditivity theorem). *Let M be a complex manifold and let φ, ψ be plurisubharmonic functions on M. Then*

$$\mathscr{I}(\varphi + \psi) \subset \mathscr{I}(\varphi) \cdot \mathscr{I}(\psi)$$

Proof. Applying the addition formula to $M_1 = M_2 = M$ and the restriction formula to $S = $ the diagonal of $M \times M$, one has $\mathscr{I}(\varphi + \psi) = \mathscr{I}((\varphi \circ \pi_1 + \psi \circ \pi_2)|_S) \subset (\mathscr{I}(\varphi \circ \pi_1 + \psi \circ \pi_2))|_S = (\pi_1^* \mathscr{I}(\varphi) \cdot \pi_2^* \mathscr{I}(\psi))|_S = \mathscr{I}(\varphi) \cdot \mathscr{I}(\psi)$.   □

Since $\mathscr{I}^t$ makes sense for any ideal $\mathscr{I} \subset \mathbb{C}\{z\}$ and any nonnegative real number t, it is natural to ask whether or not the subadditivity theorem can be generalized to

$$\mathscr{I}(t\varphi) \subset \mathscr{I}(\varphi)^t.$$

Prior to [Dm-E-L], in the study of an invariant closely related to the existence of a Kähler–Einstein metric on a complex manifold M, Demailly and Kollár [Dm-K] raised a question on the sheaf

$$\mathscr{I}_+(\varphi) := \cup_{\epsilon > 0} \mathscr{I}((1 + \epsilon)\varphi).$$

Openness Conjecture Assume that $\mathscr{I}(\varphi) = \mathscr{O}_M$. Then

$$\mathscr{I}_+(\varphi) = \mathscr{I}(\varphi).$$

It is easy to see that the above extension of subadditivity theorem will follow in the right sense if the openness conjecture is true. Besides this, the question is undoubtedly of a basic nature. In [Dm-K], quantities of central interest are the *log canonical threshold* and the *complex singularity exponent* of plurisubharmonic functions. Given a plurisubharmonic function φ, the **log canonical threshold** c_φ of φ at a point z_0 is defined as

$$c_\varphi(z_0) = \sup\{c > 0; e^{-2c\varphi} \text{ is } L^1 \text{ on a neighborhood of } z_0\} \in (0, +\infty].$$

For any compact set $K \subset M$, the **complex singularity exponent** $c_K(\varphi)$ is defined as

$$c_K(\varphi) = \sup\{c; e^{-2c\varphi} \text{ is } L^1 \text{ on a neighborhood of } K\}.$$

For the two–dimensional case, the openness conjecture was proved by Favre and Jonsson in [F-J]. For arbitrary dimension it has been reduced to a purely algebraic statement by Jonsson and Mustață (cf. [J-M]). In [Brd-2], Berndtsson solved the openness conjecture affirmatively by using symmetrization of plurisubharmonic functions. The strong openness conjecture which implies the openness conjecture was asked by Demailly in [Dm-8, Dm-9]. It is stated as follows.

Strong Openness Conjecture. For any plurisubharmonic function φ on M, one has

$$\mathscr{I}_+(\varphi) = \mathscr{I}(\varphi).$$

The strong openness conjecture was solved by Guan and Zhou [G-Z-2] by applying Theorem 3.3. A related semi-continuity theorem for the **weighted log canonical threshold**

$$c_{\varphi,f}(z_0) = \sup\{c > 0; |f|^2 e^{-2c\varphi} \text{ is } L^1 \text{ on a neighborhood of } z_0\}$$

for a holomorphic function f was obtained in [G-Z-3]. Its effective version was proved by Hiep [Hp] by combining Theorem 3.3 with a generalization of the Weierstrass division theorem due to Hironaka (cf. [H-U]).

We shall follow Hiep's proof below.

In [Hp], the main theorem is stated as follows.

Theorem 3.19. *Let f be a holomorphic function on an open set Ω in $\mathbb{C}^n$ and let $\varphi \in PSH(\Omega)(:= \{$plurisubharmonic functions on $\Omega\})$.*

(i) *("Semicontinuity theorem") Assume that $\int_{\Omega'} e^{-2c\varphi} d\lambda < \infty$ on some open subset $\Omega' \subset \Omega$ and let $z_0 \in \Omega'$. Then, for any $\psi \in PSH(\Omega')$, there exists $\delta = \delta(c, \varphi, \Omega', z_0) > 0$ such that $\|\psi - \varphi\|_{L^1(\Omega')} \le \delta$ implies $c_\psi(z_0) > c$. Moreover, as ψ converges to φ in $L^1(\Omega')$, the function $e^{-2c\psi}$ converges to $e^{-2c\varphi}$ in L^1 on every relatively compact open subset Ω'' of Ω'.*

(ii) ("Strong effective openness") Assume that $\int_{\Omega'} |f|^2 e^{-2c\varphi} d\lambda < \infty$ on some open subset $\Omega' \subset \Omega$. When $\psi \in PSH(\Omega')$ converges to φ in $L^1(\Omega')$ with $\psi \leq \varphi$, the function $|f|^2 e^{-2c\psi}$ converges to $|f|^2 e^{-2c\varphi}$ in L^1 norm on every relatively compact open subset of Ω.

Corollary 3.6 ("Strong openness"). *For any plurisubharmonic function φ on a neighborhood of a point $z_0 \in \mathbb{C}^n$, the set*

$$\{c > 0; |f|^2 e^{-2c\varphi} \text{ is } L^1 \text{ on a neighborhood of } z_0 \}$$

is an open interval $(0, c_{\varphi,f}(z_0))$.

Corollary 3.7. *("Convergence from below") If $\psi \leq \varphi$ converges to φ in a neighborhood of $z_0 \in \mathbb{C}^n$, then $c_{\psi,f}(z_0) \leq c_{\varphi,f}(z_0)$ converges to $c_{\varphi,f}(z_0)$.*

The proof is done by induction on n which is run using Hironaka's division theorem and the L^2 extension theorem (Theorem 3.3) as machinery. To state Hironaka's division theorem, we first make $\mathbb{C}\{z\}$ an ordered set. The **homogeneous lexicographical order** of monomials $z^\alpha = z_1^{\alpha_1} \cdots z_n^{\alpha_n}$ means that $z_1^{\alpha_1} \cdots z_n^{\alpha_n} < z_1^{\beta_1} \cdots z_n^{\beta_n}$ if and only if $|\alpha| = \alpha_1 + \ldots + \alpha_n < |\beta| = \beta_1 + \ldots + \beta_n$ or $|\alpha| = |\beta|$ and $\alpha_j < \beta_j$ for the first index j with $\alpha_j \neq \beta_j$. Then, for each $f = a_{\alpha^1} z^{\alpha^1} + a_{\alpha^2} z^{\alpha^2} + \ldots$ in $\mathbb{C}\{z\}$ with $a_{\alpha^j} \neq 0, j \geq 1$ and $z^{\alpha^1} < z^{\alpha^2} < \ldots$, we define the **initial coefficient, initial monomial** and **initial term** of f respectively by

$$IC(f) = a_{\alpha^1},$$

$$IM(f) = z^{\alpha^1}$$

and

$$IT(f) = a_{\alpha^1} z^{\alpha^1},$$

and the **support** of f by

$$SUPP(f) = \{z^{\alpha^1}, z^{\alpha^2}, \ldots\}.$$

For any ideal $\mathscr{I} \subset \mathbb{C}\{z\}$, $IM(\mathscr{I})$ will denote the ideal generated by $\{IM(f); f \in \mathscr{I}\}$.

Hironaka's Division Theorem (cf. [G, By, B-M-1, B-M-2, Eb]. See also [H-U].) Let $f, g_1, \ldots, g_k \in \mathbb{C}\{z\}$. Then there exist $h_1, \ldots, h_k, s \in \mathbb{C}\{z\}$ such that $f = h_1 g_1 + \ldots + h_k g_k + s$, and

$$SUPP(s) \cap \langle IM(g_1), \ldots, IM(g_k) \rangle = \varnothing,$$

where $\langle IM(g_1), \ldots, IM(g_k) \rangle$ denotes the ideal generated by the family $(IM(g_1), \ldots, IM(g_k))$.

Standard basis Let $\mathscr{I}$ be an ideal of $\mathbb{C}\{z\}$ and let $g_1, \ldots, g_k \in \mathscr{I}$ be such that $IM(\mathscr{I}) = \langle IM(g_1), \ldots, IM(g_k)\rangle$. Then, by Hironaka's division theorem it is easy to see that $g_j's$ are generators of $\mathscr{I}$. One may choose such $g_j's$ in such a way that $IM(g_1) < IM(g_2) < \cdots < IM(g_k)$, and we say that $(g_1, \ldots, g_k)$ is a **standard basis** of $\mathscr{I}$.

Now let us start the induction proof of Theorem 3.19. The idea is to apply the induction hypothesis to the restriction of f and φ to a generic hyperplane section on which one has already a better estimate, and use the L^2 extension theorem to obtain a function F with a better estimate. To derive the desired improved estimate for f from that of F, Hironaka's division theorem is applied.

First of all, the assertion is trivially true if $n = 0$. Suppose it is true for the dimension $n - 1$. Then the following is the key lemma.

Lemma 3.5. *Let* $\varphi \leq 0$ *be a plurisubharmonic function and* f *be a holomorphic function on the polydisc*

$$\Delta_R^n = \{z \in \mathbb{C}^n; |z_j| < R \text{ for all } j\}, \ R > 0$$

such that for some $c > 0$

$$\int_{\Delta_R^n} |f(z)|^2 e^{-2c\varphi(z)} d\lambda < \infty.$$

Let $\psi_\mu \leq 0, \mu \geq 1$, *be a sequence of plurisubharmonic functions on* Δ_R^n *with* $\psi_\mu \to \varphi$ *in* $L^1_{loc}(\Delta_R^n)$, *and assume that either* $f = 1$ *identically or* $\psi_\mu \leq \varphi$ *for all* $\mu \geq 1$. *Then for every* $r < R$ *and* $\epsilon \in (0, \frac{1}{2}r]$, *there exist a value* $w_n \in \Delta_\epsilon \setminus \{0\}$, *an index* μ_0, *a constant* $\tilde{c} > c$ *and a sequence of holomorphic functions* F_μ *on* Δ_r^n, $\mu \geq \mu_0$, *such that* $IM(F_\mu) \leq IM(f)$,

$$F_\mu(z) = f(z) + (z_n - w_n) \sum a_{\mu,\alpha} z^\alpha$$

with $|w_n||a_{\mu,\alpha}| \leq r^{-|\alpha|}\epsilon$ *for all* $\alpha \in \mathbb{N}^n$, *and*

$$\int_{\Delta_R^n} |F_\mu(z)|^2 e^{-2\tilde{c}\psi_\mu(z)} d\lambda \leq \frac{\epsilon^2}{|w_n|^2} < \infty$$

for all $\mu \geq \mu_0$.

Moreover, one can choose w_n *in a set of positive measure in the punctured disc* $\Delta_\epsilon \setminus \{0\}$.

Proof. By Fubini's theorem, the function

$$\int_{\Delta_R^{n-1}} |f(z', z_n)|^2 e^{-2c\varphi(z', z_n)} d\lambda_{z'}$$

in the variable z_n is integrable on Δ_R. Therefore, for any $\eta > 0$ and $\epsilon_0 > 0$, one can find $w_n \in \Delta_\eta \setminus \{0\}$ of positive measure such that

$$\int_{\Delta_R^{n-1}} |f(z', w_n)|^2 e^{-2c\varphi(z', w_n)} d\lambda_{z'} < \frac{\epsilon_0^2}{|w_n|^2}.$$

Since Theorem 3.19 is assumed to hold for $n - 1$, for any $\rho < R$ there exist $\mu_0 = \mu_0(w_n)$ and $\tilde{c} = \tilde{c}(w_n) > c$ such that

$$\int_{\Delta_\rho^{n-1}} |f(z', w_n)|^2 e^{-2\tilde{c}\psi_\mu(z', w_n)} d\lambda_{z'} < \frac{\epsilon_0^2}{|w_n|^2}$$

for all $\mu \geq \mu_0$. Hence, by extending $f(z', w_n)$ with the L^2 estimate, one has a holomorphic function F_μ on $\Delta_\rho^{n-1} \times \Delta_R$ such that $F_\mu(z', w_n) = f(z', w_n)$ for all $z' \in \Delta_\rho^{n-1}$, and

$$\int_{\Delta_\rho^{n-1} \times \Delta_R} |F_\mu(z)|^2 e^{-2\tilde{c}\psi_\mu(z)} d\lambda_z$$

$$\leq C_n R^2 \int_{\Delta_\rho^{n-1}} |f(z', w_n)|^2 e^{-2\tilde{c}\psi_\mu(z', w_n)} d\lambda_{z'}$$

$$\leq \frac{C_n R^2 \epsilon_0^2}{|w_n|^2},$$

where C_n is a constant which depends on n. Since $|F_\mu(z)|^2$ is plurisubharmonic, one has

$$|F_\mu(z)|^2 \leq \frac{1}{\pi^n(\rho - |z_1|)^2 \cdots (\rho - |z_n|)^2} \int_{\Delta_{\rho-|z_1|}(z_1) \times \ldots \times \Delta_{\rho-|z_n|}(z_n)} |F_\mu|^2 d\lambda_z$$

$$\leq \frac{C_n R^2 \epsilon_0^2}{\pi^n(\rho - |z_1|)^2 \cdots (\rho - |z_n|)^2 |w_n|^2},$$

where $\Delta_\rho(z)$ denotes the disc of radius ρ centered at z.

Hence, for any $r < R$, by taking $\rho = \frac{1}{2}(r + R)$ we infer

$$\|F_\mu\|_{L^\infty(\Delta_r^n)} \leq \frac{2^n C_n^{\frac{1}{2}} R \epsilon_0}{\pi^{\frac{n}{2}}(R - r)^n |w_n|}. \tag{3.28}$$

Let $g_\mu(z) = \sum_{\alpha \in \mathbb{N}^n} a_{\mu,\alpha} z^\alpha$ be functions on $\Delta_r^{n-1} \times \Delta_R$ satisfying

$$F_\mu(z) = f(z) + (z_n - w_n) g_\mu(z).$$

Then, by (3.28) one has

$$\|g_\mu\|_{\Delta_r^n} = \|g_\mu\|_{\Delta_r^{n-1}\times\partial\Delta_r} \leq \frac{1}{r - |w_n|}\left(\|F_\mu\|_{L^\infty(\Delta_r^n)} + \|f\|_{L^\infty(\Delta_r^n)}\right)$$

$$\leq \frac{1}{r - |w_n|}\left(\frac{2^n C_n^{\frac{1}{2}} R\epsilon_0}{\pi^{\frac{n}{2}}(R - r)^n|w_n|} + \|f\|_{L^\infty(\Delta_r^n)}\right).$$

Hence, letting $\eta \leq \epsilon_0 \leq \epsilon \leq \frac{r}{2}$, by Cauchy's estimate one has

$$|w_n||a_{\mu,\alpha}|r^{|\alpha|} \leq C'\epsilon_0$$

for some constant C' depending only on n, r, R and f. This yields the required estimates for $\epsilon_0 := C''\epsilon$ with C'' sufficiently small. As for the inequality $IM(F_\mu) \leq IM(f)$, they are achieved since one may take

$$|w_n||a_{\mu,\alpha}|r^{|\alpha|} \leq \epsilon$$

and ϵ arbitrarily small. $\square$

Before going to the proof of Theorem 3.19, Let us note that the L^1 convergence of ψ to φ implies that $\psi \to \varphi$ almost everywhere, and that the assumptions guarantee that φ and ψ are uniformly bounded on every relatively compact subset of Ω'. In particular, after shrinking Ω' and substracting constants if necessary, we may assume that $\varphi \leq 0$ on Ω. Since the L^1 topology is metrizable, we may eventually restrict ourselves to a nonpositive sequence $(\psi_\mu)_{\mu\geq 1}$ almost everywhere converging to φ in $L^1(\Omega')$. It suffices to show (i) and (ii) for some neighborhood of a given point $z_0 \in \Omega'$. For simplicity we assume $z_0 = 0$ and Δ_R^n such that $\Delta_R^n \subset \Omega'$. In this situation, $\psi(\cdot, z_n) \to \varphi(\cdot, z_n)$ in the topology of $L^1(\Delta_R^{n-1})$ for almost every $z_n \in \Delta_R$.

Proof of statement (i). By Lemma 3.5 with $f = 1$, for every $r < R$ and $\epsilon > 0$, there exist $w_n \in \Delta_\epsilon \setminus \{0\}$, $\mu_0, \tilde{c} > c$ and a sequence of holomorphic functions F_μ on Δ_r^n, $\mu \geq \mu_0$, such that $F_\mu(z) = 1 + (z_n - w_n)\sum a_{\mu,\alpha}z^\alpha$, $|w_n||a_{\mu,\alpha}|r^{-|\alpha|} \leq \epsilon$ and

$$\int_{\Delta_r^n} |F_\mu(z)|^2 e^{-2\tilde{c}\psi_\mu(z)}d\lambda_z \leq \frac{\epsilon^2}{|w_n|^2}$$

for all $\mu \geq \mu_0$.

Choosing $\epsilon \leq \frac{1}{2}$, one has $|F_\mu(0)| = |1 - w_n a_{\mu,0}| \geq \frac{1}{2}$ so that $c_{\psi_\mu}(0) \geq \tilde{c} > c$ and the first part of (i) is proved. The second assertion of (i) follows from the estimate

$$\int_{\Omega''} |e^{-2c\psi_\mu} - e^{-2c\varphi}|d\lambda_z \leq \int_{\Omega''\cap\{|\psi_\mu|\leq A\}} |e^{-2c\psi_\mu} - e^{-2c\varphi}|d\lambda_z$$

$$+ \int_{\Omega'' \cap \{|\psi_\mu| < -A\}} e^{-2c\varphi} d\lambda_z + e^{-2(\tilde{c}-c)A} \int_{\Omega'' \cap \{|\psi_\mu| < -A\}} e^{-2\tilde{c}\psi_\mu} d\lambda_z$$

since $\psi_\mu \to \varphi$ almost everywhere on Ω''.   □

Proof of statement (ii). Take $f_1, \ldots, f_k \in \mathbb{C}\{z\}$ so that $(f_1, \ldots, f_k)$ is a standard basis of $\mathscr{I}(c\varphi)_0$ with $IM(f_1) < \ldots < IM(f_k)$, and take a polydisc Δ_R^n in such a way that

$$\int_{\Delta_R^n} |f_\ell(z)|^2 e^{-2c\varphi(z)} d\lambda_z < \infty, \quad \ell = 1, \ldots, k.$$

Similarly to the above, one has $F_{\mu,\ell} \in \mathscr{I}(\tilde{c}\psi_\mu)_0 \subset \mathscr{I}(c\varphi)_0$ from f_ℓ. Namely, by Lemma 3.5, for every $r < R$ and ϵ_ℓ, there exist $w_{n,\ell} \in \Delta_{\epsilon_\ell} \setminus \{0\}$, $\mu_0 = \mu_0(w_{n,\ell})$, $\tilde{c} = \tilde{c}(w_{n,\ell}) > c$ and a sequence of holomorphic functions $F_{\mu,\ell}$ on Δ_r^n, $\mu \geq \mu_0$, such that

$$F_{\mu,\ell}(z) = f_\ell + (z_n - w_{n,\ell}) \sum a_{\mu,\ell,\alpha} z^\alpha, \quad |w_{n,\ell}||a_{\mu,\ell,\alpha}|r^{-|\alpha|} \leq \epsilon_\ell$$

and

$$\int_{\Delta_R^n} |F_{\mu,\ell}(z)|^2 e^{-2\tilde{c}\psi_\mu(z)} d\lambda \leq \frac{\epsilon_\ell^2}{|w_{n,\ell}|^2} \tag{3.29}$$

for all $\ell = 1, \ldots, k$ and $\mu \geq \mu_0$. Since $\psi_\mu \leq \varphi$ and $\tilde{c} > c$, we have $F_{\mu,\ell} \in \mathscr{I}(\tilde{c}\psi_\mu)_0 \subset \mathscr{I}(c\varphi)_0$. The next step of the proof is to modify $(F_{\mu,\ell})_{1 \leq \ell \leq k}$ into a standard basis of $\mathscr{I}(c\varphi)_0$. In virtue of (3.29) and Cauchy's estimate, by taking $\epsilon_1 \gg \epsilon_2 \gg \ldots \gg \epsilon_k$ and suitable $w_{n,\ell} \in \Delta_{\epsilon_\ell} \setminus \{0\}$, one can inductively find $F'_{\mu,\ell}$ and polynomials $P_{\mu,\ell,m}$ for $1 \leq m < \ell \leq k$ possessing uniformly bounded coefficients and degrees, such that the linear combinations

$$F'_{\mu,\ell} = F_{\mu,\ell} - \sum_{1 \leq m \leq \ell-1} P_{\mu,\ell,m} F'_{\mu,m}$$

satisfy $IM(F'_{\mu,\ell}) = IM(f_\ell)$ and $\frac{|IC(F'_{\mu,\ell})|}{|IC(f_\ell)|} \in (\frac{1}{2}, 2)$ for all ℓ and $\mu \gg 1$. In this way one finds a sequence $(F'_{\mu,1}, \ldots, F'_{\mu,k})$ of standard bases of $\mathscr{I}(c\varphi)_0$. This procedure is elementary but long, so that the reader is referred to [H] for the detail. Then, by the privileged neighborhood theorem of Siu (cf. [Siu-1]), one can find $\rho, K > 0$ with $\rho < r$ and holomorphic functions $h_{\mu,1}, \ldots, h_{\mu,k}$ on Δ_ρ^n such that

$$f = h_{\mu,1} F'_{\mu,1} + h_{\mu,2} F'_{\mu,2} + \ldots + h_{\mu,k} F'_{\mu,k} \quad on \ \Delta_\rho^n$$

and $\|h_{\mu,\ell}\|_{L^\infty(\Delta_\rho^n)} \leq K \|f\|_{L^\infty(\Delta_\rho^n)}$ for all ℓ. By (3.29) this implies a uniform bound

$$\int_{\Delta_\rho^n} |f_\ell(z)|^2 e^{-2\tilde{c}\psi_\mu(z)} d\lambda_z \leq M < \infty$$

for some $\tilde{c} > c$ and all $\mu \geq \mu_0$. The L^1 convergence of $|f|^2 e^{-2c\psi_\mu}$ to $|f|^2 e^{-2c\varphi}$ is similar to the last part of the proof of statement (i).   □

The solution of the strong openness conjecture entails a basic result on *Lelong numbers* which measures the singularity of plurisubharmonic functions. Let Ω be any open set in $\mathbb{C}^n$.

Definition 3.5. Given $\varphi \in PSH(\Omega)$ and $x \in \Omega$, the **Lelong number** of φ at x is defined as

$$\nu(\varphi, x) := \liminf_{r \to 0} \left\{ \frac{\varphi(z)}{\log r}; \|z - x\| < r \right\} \quad \left(= \lim_{r \searrow 0} \frac{\sup_{\mathbb{B}^n(x,r)} \varphi}{\log r} \geq 0 \right).$$

Example 3.5.

$$\nu\left(\log \sum_{k=1}^m |z_k|^2, 0 \right) = 2 \quad (z \in \mathbb{C}^n \text{ and } m \leq n)$$

and

$$\varphi(z) = \sum_{k=1}^\infty 2^{-k} \log \left| z - \frac{1}{k} \right| \quad (z \in \mathbb{C}) \quad \Rightarrow \nu\left(\varphi, \frac{1}{k} \right) = 2^{-k}.$$

Skoda's L^2 division theorem implies the following.

Theorem 3.20 (cf. [Sk-1]). *If $\nu(\varphi, x) < 2$ then $e^{-\varphi}$ is integrable in a neighborhood of x.*

Recently it was refined to:

Theorem 3.21 (cf. [G-Z-4](in general) and [F-J] (for $n = 2$)). *If $\nu(\varphi, x) = 2$ and $(\{z; \nu(\varphi, z) \geq 1\}, x)$ is not a germ of regular hypersurface, then $e^{-\varphi}$ is integrable on a neighborhood of x.*

Lelong numbers are related to the multiplier ideal sheaves as follows.

Theorem 3.22 (cf. [G-Z-2, G-Z-3, G-Z-5] and [B-F-J]). *For any $\varphi, \psi \in PSH(\Omega)$ and $x \in \Omega$, the following are equivalent:*

(1) For any proper holomorphic map $\pi : X \to \mathbb{C}^n$ such that $\pi|_{X \setminus \pi^{-1}(0)}$ is a local homeomorphism and for any $p \in \pi^{-1}(x)$, $\nu(\varphi \circ \pi, p) = \nu(\psi \circ \pi, p)$ holds true.
(2) $\mathscr{I}(t\varphi) = \mathscr{I}(t\psi)$ for all $t > 0$.

Chapter 4
Bergman Kernels

Abstract Applications of the L^2 method to the Bergman kernels will be discussed. Emphasis is put on the results obtained in recent decades. Among them, there are various estimates for the Bergman kernel from below on weakly pseudoconvex domains, including the solution of a long-standing conjecture of Suita by Błocki (Invent Math 193:149–158, 2013) and Guan and Zhou (Ann Math 181:1139–1208, 2015). Recently discovered variational properties due to Maitani and Yamaguchi (Math Ann 330:477–489, 2004) and Berndtsson [Brd-1] are also discussed. In a broader framework, they are describing the parameter dependence of the Bergman kernels associated to families or sequences of complex manifolds and vector bundles. Most of these new results are closely related to the L^2 extension theorems in the previous chapter. Among them, a surprise is that a variational property of the relative canonical bundles generalizing that of the Bergman kernels, which originally belongs to the theory of variation of Hodge structures, happens to imply an optimal L^2 extension theorem (cf. [Brd-L]).

4.1 Bergman Kernel and Metric

The Bergman kernel, named after Stefan Bergman (1895–1977), is by definition the reproducing kernel of the space of L^2 holomorphic n-forms on connected n-dimensional complex manifolds. Its significance in complex geometry has been gradually understood through many spectacular works in the last century. For instance, C. Fefferman [F] analyzed the boundary behavior of the Bergman kernel on strongly pseudoconvex domains with C^∞ boundary, and proved that any biholomorphic map between such bounded domains in $\mathbb{C}^n$ extends smoothly to a difffeomorphism between their closures. Recently, methods for analyzing certain generalized Bergman kernels have brought new insights into some aspects of algebraic geometry and differential geometry (cf. [Siu-7, Siu-9, Siu-10, Siu-11, Siu-12, B-Pa, Ds] and [Ma]). The purpose of this chapter is to review the results on the Bergman kernels related to such a new development.

Since it seems worthwhile to recognize the surroundings of the Bergman kernel in complex analysis and its connection to various concepts, let us briefly review some history here.

© Springer Japan 2015

T. Ohsawa, L^2 *Approaches in Several Complex Variables*, Springer Monographs in Mathematics, DOI 10.1007/978-4-431-55747-0_4

The circle division theory of C. F. Gauss (1777–1855), which was discovered on 3/30/1796 is a giant leap in mathematics and the first step towards complex geometry. In the early nineteenth century, it brought a new progress in the theory of elliptic integrals. It had been developed by L. Euler (1707–1783) and A.-M. Legendre (1752–1833) as an art of change of variables in their integration. Inspired by the work of Gauss, N.H. Abel (1802–1829) was led at first to algebraic insolvability of equations of degree 5, and subsequently discovered that the inverse functions of elliptic integrals are nothing but doubly periodic analytic functions in one complex variable, i.e. elliptic functions. He eventually reached a remarkable characterization of principal divisors in the theory of algebraic functions of one variable (Abel's theorem). The latter is now regarded as the starting point of algebraic geometry. As a generalization of Abel's theory on elliptic functions, the theory of multiply periodic functions in several variables was developed by G. Jacobi (1804–1851), K. Weierstrass (1815–1897) and B. Riemann (1826–1866). On the other hand, in spite of an important contribution of H. Poincaré (1854–1912) on normal functions and a subsequent work of S. Lefschetz (1884–1972), it was not before the appearance of the theory of harmonic integrals on Kähler manifolds by W. V. D. Hodge (1903–1975) that Abel's idea became really efficient in several variables (cf. [Ho]). This delay is, to the author's opinion, mainly because of the lack of the viewpoint of orthogonal projection in Hilbert spaces. Recall that it was only in 1899 that D. Hilbert (1862–1943) awoke Riemann's idea of Dirichlet's principle from a deep sleep (cf. [R-2]) and that the basic representation theorem of F. Riesz (1880–1956), which is often crucial in the existence proofs under orthogonality conditions, was not available until 1907. History shows that such a systematic construction in abstract mathematics emerged only after the detailed studies of orthogonal polynomials in the nineteenth century. It culminated in a general method of orthogonal projection by H. Weyl (1885–1955). Weyl's method (cf. [Wy-1]) became the analytic base of Hodge's theory, which was later combined with analytic sheaf theory by K. Kodaira [K-2, K-3] and developed into the method presented in Chap. 2. That Weyl anticipated a lot in this method is modestly suggested in [Wy-2]. Anyway, the Bergman kernel came into the picture around 1922 (cf. [Be] and [Bo]) in such a circumstance.

4.1.1 Bergman Kernels

For any set S, the set of $\mathbb{C}$-valued functions on S, simply denoted by $\mathbb{C}^S$, is naturally equipped with the structure of a complex vector space. A subspace of $\mathbb{C}^S$, say H equipped with an inner product is called a **reproducing kernel Hilbert space** if the following conditions are satisfied.

(i) H is complete with respect to the associated norm.
(ii) For any element x of S, the map $[x]$ from H to $\mathbb{C}$ defined by $[x](f) = f(x)$ is continuous.

Given a reproducing kernel Hilbert space H, Riesz's representation theorem implies that there exists uniquely a function on $S \times S$ say K_H satisfying the following:

(a) $K_H(*, y) \in H$ for any y.
(b) $(u, K_H(*, y)) = u(y)$ for any $u \in H$ and $y \in S$.

We shall say that K_H is the **reproducing kernel** of H.

Among such H and K_H, some deserve special attention when S has certain structures as manifolds or groups. For our purpose, an important example is the case where S is a complex manifold M equipped with a Hermitian metric and H is $H^{0,0}_{(2)}(M)$, the space of L^2 holomorphic functions on M.

This way of regarding $H^{0,0}_{(2)}(M)$ as a reproducing kernel Hilbert space is naturally generalized for the space of L^2 holomorphic sections of Hermitian holomorphic vector bundles over M, replacing $[x]$ in (ii) by the evaluation of the sections of E at x. Accordingly, for any Hermitian holomorphic vector bundle E over a Hermitian manifold M, or more generally for any holomorphic vector bundle with a singular fiber metric over a Hermitian manifold, $H^{0,0}_{(2)}(M, E)$ has a reproducing kernel as a section of $p_1^* E \otimes p_2^* \overline{E}$. Here p_j denotes the projection to the j-th factor of $M \times M$. Note that the spaces $H^{0,0}_{(2)}(M, E)$ are separable, so that they are isomorphic to

$$\ell^2_{\mathbb{C}} := \{(c_j)_{j=1}^\infty; c_j \in \mathbb{C} \text{ and } \sum_{j=1}^\infty |c_j|^2 < \infty\}$$

if $\dim H^{0,0}_{(2)}(M, E) = \infty$. An important special case is when E is the canonical bundle $\mathbb{K}_M$ of M. In this situation $H^{0,0}_{(2)}(M, E) = H^{n,0}_{(2)}(M)$ if M is of pure dimension n, and the inner product of the space is independent of the choices of the metric on M as long as the fiber metric of the canonical bundle is induced from the metric on M. More explicitly, $H^{n,0}_{(2)}(M)$ is equipped with a canonical inner product

$$2^{-n} i^{n^2} \int_M u \wedge \overline{v}. \tag{4.1}$$

The reproducing kernel of $H^{n,0}_{(2)}(M)$ with respect to this inner product is called the **Bergman kernel** of M, denoted by K_M instead of $K_{H^{n,0}_{(2)}(M)}$ for simplicity. A question of basic interest is how K_M detects the geometry of M. Some of the properties of K_M follow immediately from the definition. For instance, it is clear that

$$K_{M \times N}((z, s), (w, t)) = K_M(z, w) \times K_N(s, t) \tag{4.2}$$

holds up to the order of the variables.

Example 4.1. Let D be a bounded domain in $\mathbb{C}^n$, let $\Omega_z = dz_1 \wedge dz_2 \wedge \ldots \wedge dz_n$, and let $\{f_j(z)\Omega_z ; j \in \mathbb{N}\}$ be a complete orthonormal system of $H_{(2)}^{n,0}(D)$. Then

$$K_D(z, w) = \sum_{j=1}^{\infty} f_j(z)\overline{f_j(w)}\Omega_z \otimes \overline{\Omega_w}, \tag{4.3}$$

where the series converges locally uniformly on $D \times D$. Note that the boundedness assumption on D was used only to ensure that the space $H_{(2)}^{n,0}(D)$ is infinite dimensional.

A straightforward calculation yields a formula

$$K_{\mathbb{B}^n}(z, w) = (2\pi)^{-n}n!(1 - z \cdot \overline{w})^{-n-1}\Omega_z \otimes \overline{\Omega_w}, \tag{4.4}$$

where $z \cdot \overline{w} = \sum_{j=1}^{n} z_j\overline{w_j}$ and $\mathbb{B}^n = \{z \in \mathbb{C}^n; \|z\| < 1\}$ ($\|z\|^2 = z \cdot \overline{z}$).

In the situation of Example 4.1, the set $\{2^{-n/2}f_j(z) ; j \in \mathbb{N}\}$ becomes a complete orthonormal system of $H_{(2)}^{0,0}(D)$ with respect to the Lebesgue measure. The reproducing kernel $K_{H_{(2)}^{0,0}(D)}$ is called the **Bergman kernel function** of D and denoted by k_D. By an abuse of language, $k_D(z, z)$ will also be called the Bergman kernel of D. $k_D(z, z)$ is denoted by $k_D(z)$ for simplicity. Note that $k_D(z, w)$ can be recovered from $k_D(z)$, because the Taylor coefficients of $k_D(z, w)$ along the diagonal $z = w$ are all recorded in $k_D(z)$ as the Taylor coefficients in z and $\overline{z}$.

Similarly we put

$$K_M(z) = K_M(z, z) \qquad (z \in M). \tag{4.5}$$

Then $K_M(z)^{-1}$ is naturally identified with a singular fiber metric of the canonical bundle $\mathbb{K}_M$ of M whenever $K_M(z)$ does not vanish almost everywhere.

Example 4.2. If M is a complex torus $\mathbb{C}^n/\Gamma$ equipped with a metric induced from $ds_{\mathbb{C}^n}^2$, then K_M is related to the volume of M, say $Vol(\Gamma)$, by

$$K_M(z, w) = 2^{-n}Vol(\Gamma)^{-1}\Omega_z \otimes \overline{\Omega_w}. \tag{4.6}$$

From this formula, it is expected that K_M generally detects some geometric properties of M. This point will be further discussed in the following sections.

To evaluate $K_M(z)$ in more general situations, the following is most basic.

Proposition 4.1.

$$K_M(z) = \sup\{\sigma(z) \otimes \overline{\sigma(z)}; \|\sigma\|^2 = 1, \sigma \in H_{(2)}^{n,0}(M)\} \tag{4.7}$$

holds for any $z \in M$. Here the right–hand side of (4.7) is understood to be zero if $H_{(2)}^{n,0}(M) = \{0\}$.

Corollary 4.1. *For any open set $U \subset M$, $K_U(z) \geq K_M(z)$ holds for any $z \in U$.*

It is also of basic importance that $\sup\{|f(z)|^2; \|f\|^2 = 1, f \in H^{0,0}_{(2)}(D)\}$ is attained by $k_D(*, z)/\sqrt{k_D(z)}$, and that $K_{U_j}(z, w)$ converges to $K_M(z, w)$ locally uniformly on $M \times M$ if U_j increasingly converges to M.

4.1.2 The Bergman Metric

Let M_j $(j = 1, 2)$ be two complex manifolds of pure dimension n. Suppose that there exists a biholomorphic map $f : M_1 \rightarrow M_2$. Then f induces an isometry between $H^{n,0}_{(2)}(M_j)$ by the pull-back of $(n, 0)$-forms. Accordingly

$$K_{M_1} = f^* K_{M_2}. \tag{4.8}$$

In terms of the local coordinates ζ_1 around $p \in M_1$ and ζ_2 around $q \in M_2$ such that $f(p) = q$, the relation (4.8) is explicitly written as

$$k_1(\zeta_1) = k_2(f(\zeta_1))|\det(\partial\zeta_2/\partial\zeta_1)|^2, \tag{4.9}$$

where $K_{M_j}(z) = k_j(\zeta_j)\Omega_{\zeta_j} \otimes \overline{\Omega_{\zeta_j}}$. Therefore, if $K_{M_j}(z)$ are nowhere zero, the curvature forms θ_j of $K_{M_j}(z)^{-1}$ also satisfy the relation

$$\theta_1 = f^* \theta_2. \tag{4.10}$$

In particular, if $\theta_j = \sum_{\alpha,\beta=1}^{n} \theta_{j\alpha\overline{\beta}} d\zeta_j^\alpha \wedge d\overline{\zeta_j^\beta}$, and $\left(\theta_{j\alpha\overline{\beta}}\right)$ are everywhere positive definite, f is an isometry with respect to the metrics

$$\sum_{\alpha,\beta=1}^{n} \theta_{j\alpha\overline{\beta}} d\zeta_j^\alpha \otimes d\overline{\zeta_j^\beta}, \tag{4.11}$$

which are called the **Bergman metrics** of M_j. Here the notation for the metric is as a fiber metric of a holomorphic tangent bundle. The Bergman metric of a complex manifold M will be denoted by $ds^2_{M,b}$. Equation (4.10) means that biholomorphic maps preserve the Bergman metrics.

Example 4.3.

$$ds^2_{\mathbb{B}^n,b} = \left(1 - \|z\|^2\right)^{-1} ds^2_{\mathbb{C}^n} + \left(1 - \|z\|^2\right)^{-2} \partial\|z\|^2 \otimes \overline{\partial}\|z\|^2, \tag{4.12}$$

where $ds^2_{\mathbb{C}^n} = \sum_{j=1}^{n} dz_j \otimes d\overline{z}_j$. ($ds^2_{\mathbb{C}^n,b}$ does not exist.)

For a bounded domain D in $\mathbb{C}^n$, the Bergman metric $ds^2_{D,b}$ will be identified with $\partial\bar\partial \log k_D(z)$ by an abuse of notation. In terms of the notation (4.2),

$$ds^2_{D,b} = \partial\bar\partial \log k_D(z) = k_D^{-1}\partial\bar\partial k_D - k_D^{-2}\partial k_D \bar\partial k_D$$

$$= \left(\sum |f_j|^2\right)^{-2} \sum_{j\neq k} \left(\bar{f_j}\partial f_k - \bar{f_k}\partial f_j\right)\left(f_j\bar\partial f_k - f_k\bar\partial f_j\right),\qquad (4.13)$$

where $\wedge$ and $\otimes$ are confused and omitted.

Consequently, $ds^2_{D,b}$ is characterized as follows.

Proposition 4.2. $ds^2_{D,b} = \iota^* ds^2_{\mathbb{CP}^\infty}$, where $\mathbb{CP}^\infty$ is the quotient of $\ell^2_{\mathbb{C}} \setminus \{0\}$ by $\mathbb{C}^*$, $ds^2_{\mathbb{CP}^\infty}$ is induced from $ds^2_{\ell^2_{\mathbb{C}}}$, and $\iota(z) = (f_1(z) : f_2(z) : \ldots)$.

For any bounded domain $D \subset \mathbb{C}^n$, $z_0 \in D$ and $\xi \in T^{1,0}_{z_0}D$, we put

$$b(\xi) = \sup\{|\xi f|; f \in H^{0,0}_{(2)}(D), f(z_0) = 0, \|f\| = 1\}.\qquad (4.14)$$

Another immediate consequence of (4.13) is:

Proposition 4.3. In the above situation, the length of ξ with respect to $ds^2_{D,b}$ is

$$b(\xi)/\sqrt{k_D(z_0)}.\qquad (4.15)$$

4.2 The Boundary Behavior

Among the analytic properties of the Bergman kernels which reflect the geometry of complex manifolds, the boundary behavior is studied from various viewpoints. Since it is often hard to calculate the Bergman kernels explicitly, description of principal terms of their singularities and their asymptotic expansions is aimed at. It is expected that this can be achieved in terms of geometric quantities such as the Levi form of the boundary and the curvature form of the bundles. The L^2 method for the $\bar\partial$ operator is available to localize the problem. It was first applied in the case of strongly pseudoconvex domains by Hörmander [Hö-1]. To estimate the Bergman kernels from below on weakly pseudoconvex domains, the L^2 extension theorem (Theorem 3.3) is useful.

4.2.1 Localization Principle

Let the notations be as above. A basic question to be discussed here is the following: Given an open subset $U \subset M$, for which $V \subset U$, can one find a positive constant

C such that $K_U(z) \leq CK_M(z)$ holds for any $z \in V$? This is asked to understand the behavior of K_M at infinity or the boundary behavior of k_D by comparing them with those on local models. Results are described on complete Kähler manifolds in general. As we mentioned earlier, pseudoconvex domains in $\mathbb{C}^n$ are complete Kähler manifolds (cf. Corollary 2.15).

For any point z in a complex manifold M, we denote by $\mathscr{P}_z$ the set of C^∞ negative functions ϕ on $M \setminus \{z\}$ such that $e^{-\phi}$ is not integrable on any neighborhood of z. Let $U \subset M$ be an open set, let $\chi : M \to [0,1]$ be a C^∞ function such that $\chi|_{M\setminus U} = 0$, and let $V \subset \{z \in M; \chi(z) = 1\}$.

Theorem 4.1. *In the above situation, assume that M admits a complete Kähler metric and there exist $C_0 > 0$ and a bounded C^∞ function ψ on M such that one can find for every $z \in V$ a function $\phi_z \in \mathscr{P}_z$ satisfying $C_0 + \phi_z > 0$ on supp $d\chi$ and*

$$\partial\bar{\partial}(\phi_z + \psi) \geq \partial\chi\bar{\partial}\chi \tag{4.16}$$

holds on $M \setminus \{z\}$. Then there exists a constant C depending only on ψ and C_0 such that

$$K_U(z) \leq \left(1 + C\|K_U(*,z)/\sqrt{K_U(z)}\|_{\text{supp } d\chi}\right) K_M(z) \tag{4.17}$$

holds for any $z \in V$. Here $\| * \|_K$ denotes the L^2 norm over K.

Proof. Since $M \setminus \{z\}$ admits a complete Kähler metric for any fixed z, one may apply Theorem 2.14 for the trivial line bundle equipped with the fiber metric $e^{-\phi_z-\psi}$ to solve the $\bar{\partial}$–equation

$$\bar{\partial}u = \bar{\partial}(\chi \cdot K_U(*,z)/\sqrt{K_U(z)})$$

with L^2 norm estimate. Since $u(z) = 0$ by the nonintegrability of $e^{-\phi_z}$, one has a holomorphic n form

$$\chi \cdot K_U(*,z)/\sqrt{K_U(z)} - u$$

on $M \setminus \{z\}$ whose value coincides with that of $K_U(*,z)/\sqrt{K_U(z)}$ at z. By the L^2 property, it extends holomorphically to M. Thus, evaluation of the L^2 norm of $\bar{\partial}(\chi \cdot K_U(*,z)/\sqrt{K_U(z)})$ deduced from (4.16) yields (4.17). $\square$

Corollary 4.2. *Let D be a bounded pseudoconvex domain in $\mathbb{C}^n$, let $z_0 \in \partial D$ and let U_j ($j = 1,2$) be neighborhoods of z_0 in $\mathbb{C}^n$ such that $U_1 \subset\subset U_2$. Then there exists $C > 0$ such that*

$$C^{-1}k_D(z) < k_{U_2\cap D}(z) < Ck_D(z) \tag{4.18}$$

holds for any $z \in U_1 \cap D$.

Corollary 4.3. *Let D be a pseudoconvex domain in $\mathbb{C}^n$, let $z_0 \in \partial D$, and let U be a neighborhood of z_0. Suppose that ∂D is strongly pseudoconvex at z_0. Then*

$$\lim_{z \to z_0} k_{U \cap D}(z)/k_D(z) = 1. \tag{4.19}$$

In view of the proof of Theorem 4.1, it is easy to see that a similar localization principle holds for the Bergman metric.

Proposition 4.4. *In the situation of Corollary 4.2, there exists $C > 0$ such that*

$$C^{-1} b_D(\xi) < b_{U_2 \cap D}(\xi) < C b_D(\xi) \tag{4.20}$$

holds for any $\xi \in T_z^{1,0}\mathbb{C}^n$ with $z \in U_1 \cap D$.

Combining Propositions 4.3 and 4.4 with Corollary 4.2, we obtain:

Theorem 4.2. *In the situation of Corollary 4.2, there exists $C > 0$ such that*

$$C^{-1} ds_{D,b}^2 < ds_{U_2 \cap D,b}^2 < C ds_{d,b}^2 \tag{4.21}$$

holds on $U_1 \cap D$.

4.2.2 Bergman's Conjecture and Hörmander's Theorem

Let us start from a naïve observation. Let D be a bounded domain in $\mathbb{C}^n$ and let $z_0 \in \partial D$. Assume that there exist domains $D_j \subset \mathbb{C}^n (j = 1, 2)$ with $D_1 \subset D \subset D_2$ such that there exist biholomorphic authomorphisms $\alpha_j (j = 1, 2)$ of $\mathbb{C}^n$ satisfying $\alpha_1(D_1) = \mathbb{B}^n$, $\alpha_2(D_2) = \mathbb{D}^n$, $\alpha_1(z_0) \in \partial \mathbb{B}^n$ and $\alpha_2(z_0) \in \partial \mathbb{D} \times \mathbb{D}^{n-1}$. Then one can find $C > 0$ such that

$$C^{-1} \delta_D(z)^{-2} < k_D(z) < C \delta_D(z)^{-n-1} \tag{4.22}$$

if $z \in D$ and $\|z - z_0\| < \delta_D(z)$. Recall that $\delta_D(z)$ denotes the Euclidean distance from z to ∂D. In view of this, Bergman conjectured (or even asserted) that the estimate (4.22) is valid for any bounded pseudoconvex domain with C^2-smooth boundary (cf. [B-T]). Note that the existence of D_1 as above is obvious, but D_2 may not exist (cf. [K-N]). Accordingly, the estimate for k_D from below is not so straightforward. Nowadays it is known that Bergman's conjecture is true. In fact, that $C^{-1} \delta_D(z)^{-2} < k_D(z)$ follows immediately by combining Theorem 3.3 with a more or less obvious fact that it holds if $n = 1$. It was first achieved in [Oh-T-1], motivated by Hörmander's work [Hö-1] which answered Bergman's conjecture in the following way (see also [D]).

Theorem 4.3. *Let D be a pseudoconvex domain in $\mathbb{C}^n$ and let $z_0 \in \partial D$. Suppose that ∂D is strongly pseudoconvex around z_0 and let*

$$\ell(z_0) = (-1)^n \det \begin{pmatrix} \rho & \partial \rho / \partial z_j \\ \partial \rho / \partial \overline{z_k} & \partial^2 \rho / \partial z_j \partial \overline{z_k} \end{pmatrix} \Big|_{z=z_0}. \tag{4.23}$$

Here $\rho(z) = \delta_D(z)$ if $z \in D$ and $\rho(z) = -\delta_D(z)$ if $z \notin D$. Then

$$\lim_{z \to z_0} k_D(z) \delta_D(z)^{n+1} = n! \pi^{-n} \ell(z_0). \tag{4.24}$$

Proof. A direct combination of Example 4.1, Corollaries 4.1 and 4.3. □

We note that Theorem 3.4 is also available to prove Theorem 4.3 (cf. [Oh-33]).

4.2.3 Miscellanea on the Boundary Behavior

Again, let D be a pseudoconvex domain in $\mathbb{C}^n$. There are at least two types of questions related to Theorem 4.2. One is in the direction of deeper analysis on the asymptotics of $K_D(z)$ near z_0 under the strong pseudoconvexity assumption. A decisive result of this kind is the following (cf. Kerzman [Kzm] and Fefferman [F]).

Theorem 4.4. *Let $D \in \mathbb{C}^n$ be a strongly pseudoconvex domain with C^∞-smooth boundary. Then $k_D(z, w) \in C^\infty(\overline{D} \times \overline{D} \setminus \{(z, z); z \in \partial D\})$ and*

$$k_D(z) = \phi(z) \delta^{-n-1}(z) + \psi(z) \log \delta(z) \tag{4.25}$$

holds as $z \to \partial D$. Here $\overline{D}$ denotes the closure of D in $\mathbb{C}^n$ and $\phi, \psi \in C^\infty(\overline{D})$.

Another direction which we are going to describe below is less quantitative and concerns with weaker divergence properties of $k_D(z)$ and $ds_{D,b}^2$ on weakly pseudoconvex domains. One of the results motivating such studies is the following criterion for the completeness of $ds_{D,b}^2$ due to S. Kobayashi [Kb-1].

Proposition 4.5. *Suppose that $\lim_{z \to \partial D} k_D(z) = \infty$ and the set of bounded holomorphic functions on D is dense in $H_{(2)}^{0,0}(D)$. Then $ds_{D,b}^2$ is complete.*

Proof. Let $z_0 \in D$ and let $\gamma : [0, 1) \to D$ be a C^∞ curve with $\gamma(0) = z_0$ and $\gamma(t) \to \partial D$ as $t \to 1$. Then, by the assumption on $H_{(2)}^{0,0}(D)$, one can find for any $\epsilon > 0$ a bounded holomorphic function f on D such that

$$|f(z_0)|^2 = k_D(z_0) \quad \text{and} \quad \left\| f - \frac{k_D(z, z_0)}{\sqrt{k_D(z_0)}} \right\| < \epsilon.$$

136

4 Bergman Kernels

Hence, since $k_D(z)$ explodes at the boundary, one can find t_0 and an isometric embedding $\iota : D \to \mathbb{P}_{\mathbb{C}}^{\infty}(:= \mathbb{C}^N \setminus \{0\}/\mathbb{C}^*)$ such that $\iota(z_0) = (1 : 0 : 0 \ldots)$ and $\iota(\gamma(t_0)) = (0 : 1 : \ldots)$. Hence $ds_{D,b}^2$ is complete. $\qquad\square$

Kobayashi also proved that any bounded analytic polyhedron satisfies the assumptions of Proposition 4.5.

Theorem 4.5 (cf. [Kb-1]). *Let $P_1, \ldots, P_m$ be polynomials in $z = (z_1, \ldots, z_n)$ and let $D \subset \mathbb{C}^n$ be a bounded connected component of $\{z; |P_j(z)| < 1, 1 \le j \le m\}$. Then $\lim_{z \to \partial D} k_D(z) = \infty$ and $ds_{D,b}^2$ is complete.*

Applying Skoda's L^2 division theorem, P. Pflug [Pf] obtained the following.

Theorem 4.6. *Let $D \subset \mathbb{C}^n$ be a bounded pseudoconvex domain and let $z_0 \in \partial D$. Assume that there exist $\alpha > 0$ and a sequence $\{p_\mu\}_{\mu=1}^{\infty} \subset \mathbb{C}^n \setminus D$ converging to z_0 such that $\{z; \|z - p_\mu\| < \|z_0 - p_\mu\|^\alpha\} \subset \mathbb{C}^n \setminus D$ for all $\mu \in \mathbb{N}$. Then $\lim_{z \to z_0} k_D(z) = \infty$.*

Pflug's theorem suggests that $k_D(z)$ will explode along ∂D under some weak regularity assumption on ∂D. A natural class to be studied has existed for a long time in potential theory (cf. [Wn, Bou]). As a class of complex manifolds it is defined as follows.

Definition 4.1. A complex manifold M is said to be **hyperconvex** if there exists a bounded strictly plurisubharmonic exhaustion function on M.

Diederich and Fornaess [D-F-1] proved that any bounded pseudoconvex domain in $\mathbb{C}^n$ with C^2 smooth boundary is hyperconvex. Kerzman and Rosay [K-R] generalized the result to the C^1 smooth case. The following simple observation is useful.

Proposition 4.6. *M is hyperconvex if and only if there exists a strictly plurisubharmonic exhaustion function φ on M satisfying $\partial\bar\partial\varphi \ge c\partial\varphi\bar\partial\varphi$ for some positive constant c.*

Proof. Let ϕ be a bounded strictly plurisubharmonic exhaustion function on M such that $\sup_M \phi = 0$. Then

$$\partial\bar\partial(-\log(-\phi)) = \frac{\partial\bar\partial\phi}{-\phi} + \frac{\partial\phi\bar\partial\phi}{\phi^2} \ge \partial(\log(-\phi))\bar\partial(\log(-\phi)).$$

Conversely, if $\partial\bar\partial\varphi \ge c\partial\varphi\bar\partial\varphi$ for some positive constant c, one can find a bounded increasing function λ such that $\lambda(\varphi)$ is a strictly plurisubharmonic exhaustion function on M. $\qquad\square$

In virtue of the detailed study of homogeneous domains (cf. [PS]), homogeneous bounded domains are known to be hyperconvex (cf. [K-Oh]). Based on Bers's realization of Teichmüller spaces as bounded domains in $\mathbb{C}^n$, Krushkal' [Kr] showed that any finite–dimensional Teichmüller space is hyperconvex. When $\dim M = 1$, hyperconvexity of M is equivalent to the exhaustiveness of the Green function of M (cf. Proposition 3 in [Oh-16]), which can be seen easily from the definition of the Green function. Recall that the **Green function** of a Riemann surface M is by definition the maximal element of the set of continuous functions $g : M \times M \to [-\infty, 0)$ such that, for each point $w \in M$, $g(z, w)$ is subharmonic in z and, for any local coordinate ζ around w, $g(z, w) - \log |\zeta|$ is bounded on $\{z; 0 < |\zeta(z)| < 1\}$. The Green function of M will be denoted by g_M if it exists. Otherwise we put $g_M \equiv -\infty$ for the convenience of the notation.

Example 4.4.

$$g_{\mathbb{D}}(z, w) = \log \left| \frac{z - w}{1 - \bar{z}w} \right|.$$

Combining the properties of g_M with the L^2 extension theorem, one can show the following.

Theorem 4.7 (cf. [Oh-16]). *Let $D \subset \mathbb{C}^n$ be a bounded hyperconvex domain. Then* $\lim_{z \to \partial D} k_D(z) = \infty$.

For the proof, the following elementary and obvious fact is useful.

Lemma 4.1. *Let D be a bounded domain in $\mathbb{C}^n$ and let u be a bounded continuous exhaustion function on D with $\sup_D u = 0$. Then for any $\delta > 0$,*

$$\lim_{\epsilon \to 0} \left(\sup_{\zeta \in D} \quad \inf_{\ell \ni \zeta, \ell \cap \{u < -\delta\} \neq \varnothing} \int_{\ell \cap \{u > -\epsilon\}} d\lambda_\ell \right) = 0. \tag{4.26}$$

Here ℓ denotes the complex lines in $\mathbb{C}^n$ and $d\lambda_\ell$ the Lebesgue measure on ℓ.

Proof of Theorem 4.7. Let ϕ be a bounded strictly plurisubharmonic function on D with $\sup \phi = 0$. Then (4.26) holds for $u = \phi$. Moreover, as is easily seen from the definition of the Green function, for any $\delta > 0$ one can find $k > 0$ such that

$$\{z; k\phi(z) > -\epsilon\} \cap \ell \subset \{z; g_{\ell \cap D}(z, w) > -\epsilon\} \tag{4.27}$$

holds for a complex line ℓ if $\epsilon > 0$ and $\phi(w) < -\delta$. Hence, combining (4.26) and (4.27) with a well-known symmetry property $g_{\ell \cap D}(z, w) = g_{\ell \cap D}(w, z)$, we have

$$\lim_{\epsilon \to 0} \sup_{w \in \{\phi > -\epsilon\}} \quad \inf_{\ell \ni w, \ell \cap \{\phi < -\delta\} \neq \varnothing} \int_{\ell \cap g_{\ell \cap D}(z,w) < -1} d\lambda_\ell = 0 \tag{4.28}$$

for any $\delta > 0$. By (4.28) and the localization principle for the Bergman kernel, one has

$$\lim_{\epsilon \to 0} \inf_{\phi(z) > -\epsilon} \sup_{\ell \ni z} k_{\ell \cap D}(z) = \infty.$$

Hence, by the L^2 extension theorem we conclude that $\lim_{z \to \partial D} k_D(z) = \infty$ holds.

$\square$

In view of Theorem 4.7, it is natural to ask whether $ds_{M,b}^2$ is complete if M is hyperconvex. Błocki and Pflug [B-P] and Herbort [Hb] independently proved the following.

Theorem 4.8. *The Bergman metric of a bounded hyperconvex domain in $\mathbb{C}^n$ is complete.*

This was generalized by B.-Y. Chen [Ch-1]:

Theorem 4.9. *The Bergman metric of a hyperconvex manifold is complete.*

The proofs of Theorems 4.8 and 4.9 are based on Bedford and Taylor's theory of the complex Monge-Ampère operator [B-T-1, B-T-2], which is, however, beyond the scope of the present monograph.

Manipulation of the distance function with respect to the Fubini–Study metric leads to the following (cf. Ohsawa and Sibony [Oh-S]).

Theorem 4.10. *Let $D \subset \mathbb{P}^n$ be a pseudoconvex domain. Assume that ∂D is nonempty and C^2-smooth. Then D is hyperconvex.*

In fact, in the situation of the above theorem, the distance from $z \in D$ to ∂D with respect to the Fubini–Study metric, say $r(z)$, turns out to have a property that $-r(z)^\epsilon$ is strictly plurisubharmonic near ∂D for sufficiently small $\epsilon > 0$. Such a special bounded exhaustion function can be used to obtain a quantitative result.

Theorem 4.11 (cf. [D-Oh-4]). *Let $D \Subset \mathbb{C}^n$ be a pseudoconvex domain, on which there is a bounded plurisubharmonic C^∞ exhaustion function $\rho : D \to [-1, 0)$ satisfying the following estimate with suitable positive constants $C_1, C_2 > 0$.*

$$C_1^{-1}\delta_D^{C_2}(z) < -\rho(z) < C_1\delta_D^{1/C_2}(z) \tag{4.29}$$

Then there are, for any $z_0 \in D$, positive constants $c_3, c_4 > 0$ such that

$$\mathrm{dist}_D(z_0, z_1) > c_3 \log |\log (c_4\delta_D(z_1))| - 1 \tag{4.30}$$

holds for all $z_1 \in D$. Here $\mathrm{dist}_D(z_0, z_1)$ denotes the distance between z_0 and z_1 with respect to $ds_{D,b}^2$.

The proof is an application of a slight refinement of the localization principle in Theorem 4.1. Błocki [Bł-1] has improved the estimate (4.30) to

$$\mathrm{dist}_D(z_0, z_1) > \frac{\log 1/\delta_D(z_1)}{C \log |\log (c_4\delta_D(z_1))|}, \qquad C > 0. \tag{4.31}$$

The proof relies on the pluripotential theory. Whether or not

$$\mathrm{dist}_D(z_0, z_1) > \frac{\log 1/\delta_D(z_1)}{C}, \qquad C > 0. \tag{4.32}$$

holds remains an open question.

4.2.4 Comparison with a Capacity Function

Let D be a domain in $\mathbb{C}$. Then, because of the transformation formula (4.9), k_D is closely related to the theory of conformal mappings and related quantities such as capacity functions (cf. [A, Ca, S-O] and [Su-2]). In view of the L^2 method in Chap. 2, it is easy to see that $k_D \not\equiv 0$ if there exists a bounded nonconstant subharmonic function on D, or equivalently, there exists a continuous function $g : D \times D \to [-\infty, 0)$ such that the following hold for any $w \in D$:

(i) $\partial\bar\partial g(*, w) = 0$ on $D - \{w\}$.
(ii) $g(z, w) - \log|z - w|$ is bounded on a neighborhood of w.

The maximum element, say g_D, of the set of such g is nothing but the Green function of D. Accordingly, $k_D(z) \not\equiv 0$ if the Green function exists on D.

We put

$$\gamma(z)(= \gamma_D(z)) = \lim_{w \to z}(g_D(z, w) - \log|z - w|) \tag{4.33}$$

and

$$c_\beta(z)(= c_{\beta,D}(z)) = e^{\gamma(z)}. \tag{4.34}$$

γ and c_β are called the **Robin function** and the **logarithmic capacity** on D, respectively. It is straightforward that $c_\beta(z) = (1-|z|^2)^{-1}$ if $D = \mathbb{D}$. Hence $\pi k_D = c_\beta^2$.

Example 4.5. Let $\mathbb{D}(r) = \{z \in \mathbb{C}; |z| < r\}$. Then

$$g_{\mathbb{D}(r)}(z, w) = \log\left|\frac{r(z - w)}{r^2 - \bar w z}\right|, \tag{4.35}$$

$$\gamma_{\mathbb{D}(r)}(z) = \log\frac{r}{r^2 - |z|^2}, \tag{4.36}$$

$$c_{\beta,\mathbb{D}(r)}(z) = \frac{r}{r^2 - |z|^2} \tag{4.37}$$

and

$$k_{\mathbb{D}(r)}(z) = \frac{1}{\pi}\frac{r^2}{(r^2 - |z|^2)^2}. \tag{4.38}$$

Letting $\gamma \equiv -\infty$ and $c_\beta \equiv 0$ if g_D does not exist, one can say that $k_D \equiv 0$ if and only if $c_\beta \equiv 0$, as was observed by Oikawa and Sario in [S-O]. In fact, this elementary but nontrivial remark is an interpretation of Carleson's theorem on the negligible singularities of L^p holomorphic functions (cf. [Ca, §VI. Theorem 1]). The main ingredient of [Ca] is a systematic study of "thin sets" by means of capacities, Hausdorff measures, arithmetical conditions etc., dealing with the significance of these concepts to existence problems for harmonic and holomorphic functions, boundary behavior, convergence of expansions and to harmonic analysis. Based on this, Oikawa and Sario suggested comparing k_D and c_β for any domain D. The question makes sense for Riemann surfaces. Namely, using the local coordinates z and w in (4.33) and (4.34), we regard $c_{\beta,M}^2(z)dz \otimes d\bar{z}$ as a section of $T_M^{1,0} \otimes T_M^{0,1}$, so that the question is to compare $K_M(z)$ and $c_{\beta,M}^2(z)dz \otimes d\bar{z}$. By the way, the main theme of [S-O] is the study of the boundary behavior of conformal mappings aiming at applications to the classification of open (=noncompact) Riemann surfaces. At that time, a general question which attracted attention was the relation between the function spaces on a Riemann surface M and the magnitude of its *boundary*. A typical approach was to consider an extremal problem in such a way that triviality of the solution implies degeneration of certain function spaces (cf. [A-Bl]). K_M and $c_{\beta,M}$ are certainly solutions of extremal problems on M. In this context, Oikawa and Sario also asked for comparison of K_M with the **Ahlfors constant**

$$c_B(z) := \sup_{|f| \leq 1} |f'(z)|,$$

where $f \in \mathcal{O}(M)$. N. Suita (1933–2002) considered this latter question at first and solved it completely in [Su-1] with a sharp bound. After that, he proceeded to study the relation between K_M and c_β.

As for the annuli $A_r := \{r < |z| < 1\}, 0 \leq r < 1$, he proved the following in [Su-1].

Theorem 4.12. $\pi k_{A_r}(z) \geq c_{\beta,A_r}^2(z)$ *holds for all* $z \in A_r$. *The equality holds if and only if* $r = 0$.

Suita proved this by exploiting a formula of Zarankiewicz [Za] which expresses k_{A_r} in terms of the Weierstrass functions.

Conjecture 4.1 (Suita's conjecture). $\pi K_M(z) \geq c_{\beta,M}^2(z)dz \otimes d\bar{z}$ holds for any Riemann surface M. Moreover, the equality holds if and only if M is conformally (=biholomorphically) equivalent to $\mathbb{D} \setminus E$ for some E satisfying $c_{\beta,\mathbb{C}\setminus E} \equiv 0$.

In [Oh-16, Addendum], the L^2 extension theorem was applied to Suita's conjecture, and $750\pi K_M \geq c_{\beta,M}^2$ was obtained for any Riemann surface M. In 2012, Z. Błocki [Bł-2] proved:

Theorem 4.13. $\pi k_D \geq c_\beta^2$ *holds for any plane domain* D.

Błocki's proof is a refinement of a simplified variant of [Oh-T-1] by B.-Y. Chen [Ch-2]. For that, Błocki had to solve an ODE problem for two unknown functions.

The following sharpened version of Theorem 4.13 was proved by Błocki [Bł-3].

Theorem 4.14. *Let* D *be a pseudoconvex domain in* $\mathbb{C}^n$ *and let*

$$G_{D,w} = \sup\{u \in PSH(D); u < 0 \ and \ \limsup_{z \to w}(u(z) - \log|z - w|) < \infty\}.$$

Then

$$k_D(w) \geq \frac{1}{e^{2na} \mathrm{Vol}(\{G_{D,w} < -a\})}. \tag{4.39}$$

Here Vol(·) denotes the Euclidean volume.

In 2013, Q. Guan and X.-Y. Zhou [G-Z-1] proved the following, also by exploiting the solutions of an ODE problem.

Theorem 4.15. *Suita's conjecture is true for any Riemann surface.*

Theorem 4.15 is a corollary of Theorem 3.4 except for the equality criterion. There exists an intimate relation between k_D and g_D besides the above inequality:

$$\pi k_D(z, w) = 2\frac{\partial^2}{\partial z \partial \overline{w}} g_D(z, w) \qquad \text{(Bergman-Schiffer formula)} \tag{4.40}$$

and

$$\pi k_D(z) = \frac{\partial^2}{\partial z \partial \overline{z}} \gamma_D(z) \qquad \text{(Suita's formula)} \tag{4.41}$$

F. Maitani and H. Yamaguchi [M-Y] have exploited (4.41) to obtain an interesting variational property of k_D. According to a striking work of B. Berndtsson and L. Lempert [Brd-L], which was inspired by Theorem 4.14 (according to Lempert), a generalization of the Maitani–Yamaguchi theorem implies Theorem 4.13. These materials will be discussed in Sects. 4.4.2, 4.4.3 and 4.4.4.

4.3 Sequences of Bergman Kernels

Let M_μ ($\mu \in \mathbb{N}$) be a sequence of Hermitian manifolds, let E_μ be holomorphic vector bundles over M_μ, and let h_μ be singular fiber metrics of E_μ. Then, the behavior of the associated sequence of reproducing kernels $K_{H^{0,0}_{(2)}(M_\mu, E_\mu)}$ is expected to reflect that of (M_μ, E_μ, h_μ). Some instances of results in this direction will be presented below. $K_{H^{0,0}_{(2)}(M_\mu, E_\mu)}$ and its restriction to the diagonal will also be called the Bergman kernels.

4.3.1 Weighted Sequences of Bergman Kernels

Let (M, ω) be a complete Kähler manifold, let ϕ be a nonnegative C^∞ plurisubharmonic function on M and let M_0 be the interior of $\{z \in M; \phi(z) = 0\}$. For the sequence $K_{H^{n,0}_{(2)}(M, e^{-m\phi})}$, ($m \in \mathbb{N}$), the following is straightforward by the L^2 method.

Proposition 4.7. *On $M_0 \times M_0$, $K_{H^{n,0}_{(2)}(M,e^{-m\phi})}$ locally uniformly converges to K_{M_0} as $m \to \infty$.*

Let (L, b) be a positive line bundle over a connected compact Kähler manifold (M, ω) of dimension n. The behavior of the sequence $K_{H^{0,0}_{(2)}((M,\omega),(L^\mu,b^\mu))}$ is related to the existence of certain extremal metrics on M as was suggested by S.-T. Yau in [Yau-2]. The first result indicating this relationship was shown by G. Tian [Ti]. For simplicity we put $K_{M,\mu}(z, w) = K_{H^{0,0}_{(2)}((M,\omega),(L^\mu,b^\mu))}(z, w)$ and $K_{M,\mu}(z) = K_{M,\mu}(z, z)$. Tian proved:

Theorem 4.16. $\lim_{\mu\to\infty} K_{M,\mu}(z)^{1/\mu} = b(z)^{-1}$ *holds for any $z \in M$.*

Proof. Let $z_0 \in M$ be any point. Let z be a local coordinate around z_0 such that $\omega = i \sum_{j=1}^n dz_j \wedge d\overline{z_j} + O(\|z\|^2)$, and let ζ be a fiber coordinate of L over a neighborhood U of z_0 such that $b(z, \zeta) = |\zeta|^2 + O(\|z\|^2)$. Let s be a C^∞ section of L which is identically equal to 1 with respect to ζ on a neighborhood of z_0 and $\equiv 0$ outside U. Put $v = \overline{\partial}\left(\chi(\sqrt{\mu}\|z\|)s^\mu\right)$, where χ is a C^∞ real–valued function on $\mathbb{R}$ such that $supp\chi \subset [-2, 2]$ and $\chi \equiv 1$ on $[-1, 1]$. Then, by Theorem 2.14, one can solve the $\overline{\partial}$-equation $\overline{\partial}u = v$ with a side condition $u(z_0) = 0$ and with an L^2 estimate $\|u\|^2 \le C$, where C is a constant independent of μ. Hence one has an element $\chi(\sqrt{\mu}\|z\|)s^\mu - u$ of $H^{0,0}_{(2)}((M, \omega), (L^\mu, b^\mu))$ approximating $K_{M,\mu}(z_0)^{1/\mu}$ in the desired way. □

Tian proved moreover that $K_{M,\mu}(z)^{1/\mu}$ converges to $b(z)$ in the C^2-topology. As a result, $1/\mu$ times the curvature form of $K_{M,\mu}(z)^{-1}$ converges to Θ_b. (Recall that Θ_b denotes the curvature form of b.)

Later, D. Catlin [Ct] and S. Zelditch [Ze] independently proved the following.

Theorem 4.17. *In the above situation, assume moreover that $\omega = i\Theta_b$. Then there exist C^∞ functions $a_m(m = 0, 1, 2, \ldots)$ on M such that the asymptotic expansion*

$$K_{M,\mu}(z) \sim a_0(z)\mu^n + a_1(z)\mu^{n-1} + a_2(z)\mu^{n-2} + \cdots \qquad (4.42)$$

holds with $a_0(z) = 1$. Here $L^\mu \otimes \overline{L^\mu}$ is identified with the trivial bundle by the fiber metric b.

In [Ct] and [Ze], an asymptotic formula of Boutet de Monvel and Sjöstrand for the boundary behavior of the Bergman kernel, which is similar to (4.42), was used. It may be worthwhile to note that the above proof of Tian's theorem can be refined to give an elementary proof of Theorem 4.17. (See [B-B-S].)

Apparently there exists a parallelism between Theorems 4.3, 4.7 and Theorems 4.16, 4.17, the counterpart of 4.3 (resp. 4.7) being 4.16 (resp. 4.17). Strong pseudoconvexity of ∂D corresponds to the (strict) positivity of (L, b). Accordingly, it is natural to expect that Theorem 4.16 can be extended as a convergence theorem for $K_{M,\mu}(z)^{1/\mu}$ under weaker positivity assumptions. Such an instance is an approximation theorem of Demailly to be explained below.

4.3.2 Demailly's Approximation Theorem

Let D be a domain in $\mathbb{C}^n$ and let $\phi(z)$ be a plurisubharmonic function on D ($\phi \in PSH(D)$). Recall that $\phi(z)$ can be locally approximated from above by C^∞ plurisubharmonic functions (cf. Sect. 1.2.1). It was shown by Bremermann [Brm] that any $\phi \in PSH(D)$ can be approximated on compact subsets of D by linear combinations of $\log|f|$ for $f \in \mathcal{O}(D)$ as long as D is pseudoconvex. This is because the domain $\{(z, w); z \in D$ and $|w| < e^{-\phi(z)}\}$ becomes pseudoconvex and therefore holomorphically convex by the solution of the Levi problem. Demailly [Dm-6] has shown a more quantitative approximation theorem for plurisubharmonic functions in the spirit of the Bergman kernels $k_{D,m}(z) := K_{H^{0,0}_{(2)}(D,e^{-m\phi})}$.

Theorem 4.18. *Let D be a bounded pseudoconvex domain in $\mathbb{C}^n$ and let $\phi \in PSH(D)$. Then there are constants $C_1, C_2 > 0$ such that*

$$\phi(z) - \frac{C_1}{m} \le \frac{1}{m}\log k_{D,m}(z) \le \sup_{|\zeta-z|<r} \phi(\zeta) + \frac{1}{m}\log\frac{C_2}{r^n}. \tag{4.43}$$

for every $z \in D$ and $r < \delta_D(z)$.

Proof. Let $z_0 \in D$, let $f \in \mathcal{O}(D)$, and let $r < \delta_D(z_0)$. Then the mean value inequality applied to $|f|^2$ implies

$$|f(z_0)|^2 \le \frac{n!}{\pi^n r^{2n}} \int_{\|z-z_0\|<r} |f(z)|^2 d\lambda(z) \tag{4.44}$$

$$\le \frac{n!}{\pi^n r^{2n}} \exp\left(\sup_{\|z-z_0\|<r} m\phi(z)\right) \int_D |f|^2 e^{-m\phi} d\lambda, \tag{4.45}$$

from which the second inequality in (4.43) is easy to see. The first inequality is an immediate consequence of an estimate for $k_{D,m}(z_0)$ from below by the L^2 extension theorem (Theorem 3.3) applied to (D, z_0) with respect to $(d\lambda, e^{-m\phi})$. $\qquad\square$

Corollary 4.4 (Siu's theorem). *Let D and ϕ be as above. Then, given a positive number c, the set*

$$E_c(\phi) := \{z \in D; \nu(\phi, z) \ge c\} \qquad (\text{recall Definition 3.5})$$

is an analytic set in D.

Proof. From Theorem 4.18,

$$E_c(\phi) = \bigcap_{m\ge 1} E_{c-\frac{n}{m}}(\log k_{D,m}(z)).$$

Since

$$E_{c-\frac{n}{m}}(\log k_{D,m}(z)) = \{z; |\alpha| < mc-n \text{ implies } f^{(\alpha)}(z) = 0 \text{ for any } f \in H^{0,0}_{(2)}(D, e^{-m\phi})\},$$

it follows that $E_{c-\frac{n}{m}}$ are analytic sets, and so is the intersection $E_c(\phi)$ of these sets.

□

Corollary 4.4 was first proved in [Siu-4] by exploiting the classical technique in Chap. 2. For more materials related to the Lelong number, e.g. positive currents, the reader is referred to [Dm-9].

4.3.3 Towering Bergman Kernels

Up to Theorem 4.18, Bergman kernels on a fixed manifold were considered. The method can be naturally extended to study the Bergman kernels on a family of different base manifolds. A particularly interesting situation arises when the group actions are involved.

By a **tower** of complex manifolds, we shall mean a sequence of complex manifolds $M_j(j = 1, 2, \ldots)$ such that $M_j = M/\Gamma_j$ for some decreasing sequence of discrete subgroups Γ_j of the group $Aut(M)$ of biholomorphic automorphisms of M, acting on M properly discontinuously without fixed points such that $\cap_{j=1}^{\infty}\Gamma_j = \{id_M\}$ and $[\Gamma_1 : \Gamma_j] < \infty$ for all j. We put $M_\infty = M$. A tower $\{M_j\}_{j=1}^{\infty}$ is said to be **normal** if Γ_j is a normal subgroup of Γ_1 for all j. Given a tower $\{M_j\}$, one has natural projections $\pi_{k,j} : M_k \to M_j(j < k)$ which are Galois coverings if $\{M_j\}$ is normal.

In this setting, a natural question is whether or not the sequence $\pi_{\infty,j}^* K_{M_j}(z)$ converges to $K_{M_\infty}(z)$.

J. A. Rhodes [Rh] proved the following.

Theorem 4.19. *If $\{M_j\}$ is a normal tower with $M_\infty = \mathbb{D}$ such that M_1 is compact, $\pi_{\infty,j}^* K_{M_j}(z)$ converges to $K_{M_\infty}(z)$ locally uniformly.*

Sketch of proof. By the $Aut(\mathbb{D})$ invariance of $ds_{\mathbb{D}}^2$, it follows from Proposition 4.1 and Cauchy's estimate that the sequence $\pi_{\infty,j}^* K_{M_j}(z)$ is equicontinuous with respect to $ds_{\mathbb{D}}^2$. By the assumption $\cap_{j=1}^{\infty}\Gamma_j = \{id_M\}$, the inequality

$$\limsup_{j\to\infty} \pi_{\infty,j}^* K_{M_j}(z) \leq K_{M_\infty}(z) \tag{4.46}$$

is obvious. On the other hand, since M_j are compact for all $j \neq \infty$, by the Gauss–Bonnet formula and (4.4) one knows

$$\int_{M_j} K_{M_j}(z) = \int_{M_j} K_{M_\infty}(z). \tag{4.47}$$

Here $K_{M_\infty}(z)$ is identified with a form on M_j by the invariance under Γ_j. Since $\pi_{\infty,j}^* K_{M_j}(z)$ are equicontinuous, combining (4.45) and (4.46) with the normality of the tower, the desired convergence is obtained. □

In [Oh-25], an example of a non-normal tower of compact Riemann surfaces is given for which $\pi_{\infty,j}^* K_{M_j}(z)$ does not converge to $K_{M_\infty}(z)$.

4.4 Parameter Dependence

Given a continuous family of n-dimensional complex manifolds M_t and Hermitian holomorphic vector bundles E_t over M_t, the dependence of $K_{H_{(2)}^{n,0}(M_t,E_t)}$ on t will be discussed.

4.4.1 Stability Theorems

Let M be a complex manifold, let $\pi : M \to \mathbb{D}$ be a surjective holomorphic map with smooth fibers, and let (E, h) be a holomorphic Hermitian vector bundle over M. We put $M_t = \pi^{-1}(t)$, $E_t = E|_{M_t}$ and $K_{M,E,t} = K_{H_{(2)}^{n,0}(M_t,E_t)}$, where $n = \dim M - 1$.

By Proposition 4.1 and Cauchy's estimate, it is easy to see that $K_{M,E,t}(z)$ is upper semicontinuous on $\mathbb{D} \times M$. If π is proper, it is also immediate that $K_{M,E,t}(z)$ is continuous if and only if $\dim H^{n,0}(M_t, E_t)$ is constant. It is the case if E_t are Nakano positive as is easily seen from Theorem 2.20. If M_t are Kählerian, that $\dim H^{n,0}(M_t)$ does not depend on t follows from the Hodge decomposition (cf. Theorem 2.33) and the upper semicontinuity of $\dim H^{p,q}(M_t)$ in t. Moreover, as we have noted in Sect. 3.1.4, the invariance of $\dim H^{n,0}(M_t, E_t)$ is also true if M is holomorphically embeddable into some $\mathbb{CP}^N$ and $E_t \simeq \mathbb{K}_{M_t}^m$ for some $m \in \mathbb{N}$.

For the case of Nakano positive bundles, one can also deduce from Theorem 2.14 a continuity result for nonproper π.

Theorem 4.20. *Let $\pi : M \to \mathbb{D}$ be a surjective holomorphic map with smooth fibers, and let (E, h) be a Nakano positive vector bundle over M. Assume that there exists a complete Kähler metric on M_t for every t depending continuously on t and there exists a Kähler metric ds^2 on M and a constant $c > 0$ such that the least eigenvalue of the Nakano form of (E, h) with respect to ds^2 is everywhere $\geq c$. Then $K_{M,E,t}(z)$ is continuous.*

Concerning the proof of Theorem 4.20, the reader is referred to [D-Oh-3] for the argument of solving the $\bar\partial$-equation continuously in t by applying Theorem 2.14.

Remark 4.1. It might be interesting if one can find a natural condition for the continuity of $K_{M,\omega_{M_t}^m}$, where the fiber metric of $\mathbb{K}_{M_t}^m$ is induced from the Bergman kernel on M_t.

As for the Bergman metrics $ds_{M_t}^2$, a criterion for the continuity can be described similarly to that above.

Theorem 4.21. *Let M, π, ds_t^2 be as in Theorem 4.19. Assume moreover that there exists a bounded strictly plurisubharmonic function on M. Then the family of Bergman metrics $ds_{M_t}^2$ is continuous on $M \times \mathbb{D}$.*

Continuity of the derivatives of the Bergman kernels can be analyzed similarly as long as the directions of derivatives are tangent to M_t (cf. [G-K]). As for the smoothness with respect to t, not many facts seem to be known except for a recent work of X.-H. Gong and K.-T. Kim [G-Km].

4.4.2 Maitani–Yamaguchi Theorem

F. Maitani and H. Yamaguchi [M-Y] observed from

$$\log k_{\mathbb{D}(e^{-\phi(t)})}(z) = -\log \pi + 2\phi(t) + 2 \sum_{j=1}^{\infty} \frac{(e^{\phi(t)}|z|)^{2j}}{j}, \qquad (4.48)$$

which is known to hold for any upper semicontinuous function $\phi = \phi(t)$ on $\mathbb{D}$ by (4.38), that $\log k_{\mathbb{D}(e^{-\phi(t)})}(z)$ is plurisubharmonic in (t, z) if and only if $\phi(t)$ is subharmonic in t. They generalized this to the following.

Theorem 4.22. *Let $p : \mathscr{D} \to \mathbb{D} \times \mathbb{C}$ be a Riemann domain, let $p_1 : \mathbb{D} \times \mathbb{C} \to \mathbb{D}$ be the projection to the first factor, and let $\mathscr{D}_t = p^{-1}(\{t\} \times \mathbb{C})$. Assume that $\mathscr{D}$ is pseudoconvex. Then $\log k_{\mathscr{D}_t}(z)$ $(z \in \mathscr{D}_t)$ is plurisubharmonic on $\mathscr{D}$.*

Let us present below only a sketchy account on the proof of Theorem 4.22, since the result was later generalized by Berndtsson [Brd-1] and Guan and Zhou [G-Z-1] by completely different methods. Their proofs will be discussed later in detail.

Sketch of proof. Since the assertion is local in t, one may assume that $\partial \mathscr{D}$ is smooth and $\mathscr{D}_t$ are domains with real analytic smooth boundary. Let Φ be a defining function of $\partial \mathscr{D}$. We put $g(t, z, w) = g_{\mathscr{D}_t}(z, w)$ and $\gamma(t, z) = \gamma_{\mathscr{D}_t}(z)$.

Since

$$g(t, z, w) = \log \frac{1}{|z - w|} + \gamma(t, z) + h(t, z, w), \qquad (4.49)$$

where h is harmonic in z in a neighborhood of $w \in \mathscr{D}_t$ and $h(t, w, w) = 0$ for all t, $\partial \gamma(t, z)/\partial t$ and $\partial^2 \gamma(t, z)/\partial t \partial \bar{t}$ are harmonic on $\mathscr{D}_t$. Hence, by a generalized Poisson formula on $\mathscr{D}_t$, one obtains

$$\frac{\partial \gamma(t, w)}{\partial t} = \frac{1}{\pi} \int_{\partial \mathscr{D}_t} \left(\frac{\partial \Phi}{\partial t} \bigg/ \left| \frac{\partial \Phi}{\partial z} \right| \right) \left| \frac{\partial g(t, z, w)}{\partial z} \right|^2 ds \qquad (4.50)$$

Here ds denotes the arc length element of $\partial\mathscr{D}_t$. Similarly, modifying the contour integral by Stokes' formula, one has

$$\frac{\partial^2 \gamma(t,w)}{\partial t \partial \bar{t}} = \frac{1}{\pi} \int_{\partial\mathscr{D}_t} L(t,z) \left| \frac{\partial g(t,z,w)}{\partial z} \right|^2 ds + \frac{4}{\pi} \int_{\mathscr{D}_t} \left| \frac{\partial^2 g(t,z,w)}{\partial \bar{t} \partial z} \right|^2 d\lambda, \quad (4.51)$$

where

$$L(t,z) = \left(\frac{\partial^2 \Phi}{\partial t \partial \bar{t}} \left| \frac{\partial \Phi}{\partial z} \right|^2 - 2\mathrm{Re}\left\{ \frac{\partial^2 \Phi}{\partial \bar{t} \partial z} \frac{\partial \Phi}{\partial t} \frac{\partial \Phi}{\partial \bar{z}} \right\} + \left| \frac{\partial \Phi}{\partial t} \right|^2 \frac{\partial^2 \Phi}{\partial z \partial \bar{z}} \right) \Big/ \left| \frac{\partial \Phi}{\partial z} \right|^3.$$

$$(4.52)$$

Let

$$\mathscr{L}(t,z,w) = \frac{2}{\pi} \frac{\partial g(t,z,w)}{\partial z \partial w}$$

and

$$\mathscr{K}(t,z,w) = \frac{2}{\pi} \frac{\partial g(t,z,w)}{\partial z \partial \bar{w}}.$$

Then, combining (4.49) and (4.50) with the Bergman-Schiffer and Suita formulas (cf. (4.40) and (4.41)), one has

$$\frac{\partial^2 k_{\mathscr{D}_t}(w)}{\partial t \partial \bar{t}} = \frac{1}{4} \int_{\partial\mathscr{D}_t} L(t,z) \left(|\mathscr{L}(t,z,w)|^2 + |\mathscr{K}(t,z,w)|^2 \right) ds$$

$$+ \int_{\mathscr{D}_t} \left(\left| \frac{\partial \mathscr{L}(t,z,w)}{\partial \bar{t}} \right|^2 + \left| \frac{\partial \mathscr{K}(t,z,w)}{\partial \bar{t}} \right|^2 \right) d\lambda. \quad (4.53)$$

(The computation is actually quite involved.)

Hence, since $L(t,z)$ is nonnegative by the pseudoconvexity assumption, $k_{\mathscr{D}_t(z)}$ is subharmonic in t for any fixed z. By a holomorphic coordinate transformation of the form $(t,z) \to (t, w + f(t)(z - w))$ and by using the subharmonicity in t of the transformed Bergman kernels for fixed z, one deduces that $\log k_{\mathscr{D}_t(z)}$ is subharmonic in t for any z. Similarly one sees the subharmonicity of $\log k_{\mathscr{D}_t(z)}$ along any local holomorphic sections of $\mathscr{D} \to \mathbb{D}$. Therefore $\log k_{\mathscr{D}_t(z)}$ is plurisubharmonic in (t,z). $\qquad\square$

Formulas (4.49) and (4.50) can be regarded as variants of the corresponding formulas in [Y-3] and [L-Y-1] for the Robin functions on higher–dimensional domains. Although $\frac{\partial \gamma(t,w)}{\partial t}$ and $\frac{\partial^2 \gamma(t,w)}{\partial t \partial \bar{t}}$ had been studied in [Y-1, Y-2], motivated by a classification theory of entire functions (cf. [Ni-1]), their relation to the Bergman kernel was made explicit only in [M-Y] as an application of formulas of Schiffer and

Suita. So, [M-Y] has the nature of a side remark in the theory of Robin functions. As a further development of Yamaguchi's theory, see [Ha] for instance.

Unfortunately, the higher–dimensional variational formulas in [L-Y-1] are difficult to apply to study the Bergman kernels since there is no precise analogue of (4.40) and (4.41) for $n > 1$. In this situation, B. Berndtsson came up with a completely new method of generalizing Theorem 4.22 to higher dimensions by making an observation that the plurisubharmonicity of $\log k_{\mathscr{D}_t(z)}$ in (t, z) is a consequence of Nakano semipositivity of a Hilbertian bundle over $\mathbb{D}$. The purpose of Sect. 4.4.3 is to review this work after [Brd-1].

4.4.3 Berndtsson's Method

Let $\pi : M \to \mathbb{D}$ be as in Sect. 4.4.1. A natural question after Theorem 4.22 is whether or not $K_{M_t}(z)^{-1}$ is a singular fiber metric of the bundle $\mathbb{K}_{M/\mathbb{D}} := \mathbb{K}_M \otimes \pi^* \mathbb{K}_{\mathbb{D}}^{-1}$ with positive curvature current if M is a Stein manifold. This is indeed the case as explained below after [Brd-1].

Let Ω be a bounded domain in a complex manifold M_0. We assume that Ω admits a complete Kähler metric and a bounded plurisubharmonic function which is strictly plurisubharmonic at some point. Let ϕ be any C^∞ plurisubharmonic function on a neighborhood of $\overline{\mathbb{D} \times \Omega}$ in $\mathbb{D} \times M_0$. For each $t \in \mathbb{D}$, put $\phi^t(*) = \phi(t, *)$ and denote by A_t^2 the space $H_{(2)}^{n,0}(\Omega)$ with respect to the norm $\|h\|^2 = \|h\|_t^2 = i^{n^2} \int_\Omega h \wedge \overline{h} e^{-\phi^t}$. Let E be the vector bundle over $\mathbb{D}$ of infinite rank with fiber $E_t = A_t^2$.

Theorem 4.23. *If ϕ is plurisubharmonic, then the Hermitian bundle $(E, \| * \|_t)$ is semipositive. If ϕ is strictly plurisubharmonic, then $(E, \| * \|_t)$ is positive.*

Proof. Let L_t^2 denote the space $L_{(2)}^{n,0}(\Omega, e^{-\phi^t})$ consisting of L^2 $(n,0)$-forms on Ω equipped with the norm $\| * \|_t$. By F we denote the vector bundle with fiber L_t^2, a trivial bundle with a possibly nontrivial metric. Clearly, just as in the finite rank case, the curvature form of this Hermitian bundle F is given by the operator of multiplication by $(\partial^2 \phi / \partial t \partial \overline{t}) \, dt \wedge d\overline{t}$. To get an expression of the curvature form $\Theta_E dt \wedge d\overline{t}$ of $(E, \| * \|_t)$, we apply the Griffiths formula (3.16) and obtain

$$\left(\frac{\partial^2 \phi}{\partial t \partial \overline{t}} u, v \right) = \left(\pi_\perp \left(\frac{\partial \phi}{\partial t} u \right), \pi_\perp \left(\frac{\partial \phi}{\partial t} v \right) \right) + (\Theta_E u, v). \tag{4.54}$$

for any u and $v \in L_t^2$ lying in the domain of Θ_E. Here $\pi_\perp$ denotes the orthogonal projection to the orthogonal complement of E. (u and v do not depend on t.) Hence, in order to show the semipositivity of E, it suffices to show that

$$\left\| \pi_\perp \left(\frac{\partial \phi}{\partial t} u \right) \right\|^2 \leq \left(\frac{\partial^2 \phi}{\partial t \partial \overline{t}} u, u \right) \tag{4.55}$$

holds for any $u \in A_t^2$. However, noticing that

$$\bar{\partial}_z \left(\frac{\partial \phi}{\partial t} u \right) = \sum_{j=1}^{n} \frac{\partial^2 \phi}{\partial t \partial \overline{z_j}} d\overline{z_j} \wedge u,$$

and so that $\pi_\perp \left(\frac{\partial \phi}{\partial t} u \right)$ is the L^2 minimal solution of

$$\bar{\partial}_z w = \sum_{j=1}^{n} \frac{\partial^2 \phi}{\partial t \partial \overline{z_j}} d\overline{z_j} \wedge u,$$

one can easily see from (2.17) that (4.54) is valid for any $u \in A_t^2$ as long as ϕ is plurisubharmonic. Positivity of E for strictly plurisubharmonic ϕ is similar. $\square$

Let $K_{\Omega, \phi_t}(z) = K_{A_t^2}(z, z)$.

Corollary 4.5. *In the above situation, $K_{\Omega, \phi_t}(z)^{-1}$ is a singular fiber metric of $\mathbb{K}_{(\mathbb{D} \times \Omega)/\mathbb{D}}$ with positive curvature current.*

Let ψ be a C^∞ strictly plurisubharmonic function on $\overline{\mathbb{D} \times \Omega}$ and let $D_c = \{(t, z) \in \mathbb{D} \times \Omega; \psi(t, z) < c\}$. By Proposition 4.7, Corollary 4.5 is immediately extended to the following more general positivity assertion.

Proposition 4.8. *The curvature form of $K_{D_{c,t}}(z)^{-1}$ is positive on D_c.*

Furthermore, by a continuity argument, one can immediately conclude from Corollary 4.5 that the following generalization of Theorem 4.22 holds.

Theorem 4.24. *Let M be a connected Stein manifold and let $\pi : M \to \mathbb{D}$ be a surjective holomorphic map with smooth fibers. If $K_{M_t} \not\equiv 0$, then $K_{M_t}(z)^{-1}$ $(z \in M_t)$ is a singular fiber metric of the bundle $\mathbb{K}_{M/\mathbb{D}}$ with positive curvature current.*

As for the positivity of the curvature on higher–dimensional parameter spaces, the above proof of Theorem 4.23 can be naturally extended to the $H_{(2)}^{n,0}(\Omega)$-bundles $(E, \| * \|_t)$ over the polydiscs $\mathbb{D}^m$ for a C^∞ function ϕ on a neighborhood of $\overline{\mathbb{D}^m \times \Omega}$ to obtain the following.

Theorem 4.25. *If ϕ is plurisubharmonic (resp. strictly plurisubharmonic), then $(E, \| * \|_t)$ is Nakano semipositive (resp. Nakano positive).*

Theorem 4.25 is naturally generalized to the following (see also [Brd-1]).

Theorem 4.26. *Let $f : X \to U$ be a Kähler fibration with compact fibers over an open set U in $\mathbb{C}^m$. Let L be a (semi-)positive line bundle over X and let $E \to U$ be the vector bundle with $\mathcal{O}(E) = f_* \mathcal{O}(L \otimes \mathbb{K}_X \otimes f^* \mathbb{K}_U^{-1})$ such that E is equipped with the canonical fiber metric induced from the L^2 inner product on the fibers of f. Then E is Nakano (semi-)positive.*

Here, "f is a Kähler fibration" means that f is differentiably a locally trivial fibration and every fiber of f admits a neighborhood carrying a Kähler metric. Berndtsson's proof is simple enough and theoretically important because it clarifies a link between a variational problem and the optimality of the L^2-estimate (2.16) for the $\bar\partial$-operator originally due to Kodaira and Hörmander. It must be noted that Theorem 4.26 generalizes also Fujita's semipositivity theorem which was discovered in a context of the classification theory of projective algebraic varieties. Yet there is still another approach to Theorem 4.22 which was recently discovered by Q. Guan and X.-Y. Zhou [G-Z-1]. This method is even simpler than Berndtsson's, once the L^2 extension theorem with optimal constant is admitted. For the detail see the next paragraph.

4.4.4 Guan–Zhou Method

Let the notation be as above. Guan and Zhou's approach to Berndtsson's theorem (Theorem 4.24) starts from an observation that the L^2 extension theorem can be understood as a sub-mean-value property of the fiberwise Bergman kernels, as one can see from Proposition 4.1. Actually, to prove the following, it suffices to make this remark more precise.

Theorem 4.27 (cf. [G-Z-1]). *Let M be a connected complex manifold and let p be a holomorphic map with smooth fibers from $M \setminus X$ is Stein and $p^{-1}(0) \cap X$ is nowhere dense in $p^{-1}(0)$. Then, either $K_{M_t}(z) \equiv 0$ or $K_{M_t}(z)^{-1}$ is a singular fiber metric of the bundle $\mathbb{K}_{M/\mathbb{D}}$ with positive curvature current.*

Proof. Let z_0 be any point of M_0 and let (t, z) be a local coordinate of M around $(0, z_0)$. In terms of (t, z), the Bergman kernels $K_{M_t}(z)$ are expressed as

$$K_{M_t}(z) = 2^{-n}k(t,z)dz_1 \wedge dz_2 \wedge \ldots \wedge dz_n \otimes d\overline{z_1} \wedge d\overline{z_2} \wedge \ldots \wedge d\overline{z_n} \quad (4.56)$$

To prove the assertion, it suffices to show that the average of $\log k(t, z_0)$ over $|t| < r$ is $\geq \log k(0, z_0)$ for such a coordinate (t, z). For that, let us put

$$u_0 = \frac{K_{M_0}(z, z_0)}{\sqrt{k(0, z_0)}d\overline{z_1} \wedge d\overline{z_2} \wedge \ldots \wedge d\overline{z_n}}. \quad (4.57)$$

By the L^2 extension theorem, if $0 < r \leq 1$ there exists a holomorphic section $\tilde{u}_r$ of $\omega_{M/\mathbb{D}}|_{p^{-1}(\{|t|<r\})}$ such that

$$u_0 = \tilde{u}_r|_{M_0} \quad \text{and} \quad \|\tilde{u}_r\|^2 \leq \pi r^2. \quad (4.58)$$

On the other hand, by Proposition 4.1,

$$K_{M_t}(z) \geq \frac{\tilde{u}_r(z) \otimes \overline{\tilde{u}_r(z)}}{\|\tilde{u}_r\|_t^2} \quad (4.59)$$

if $|t| < r$. Here the right hand side is to be read as 0 if $\|\tilde{u}_r\|_t = \infty$. Note that the latter part of (4.57) means

$$1 \geq \frac{1}{\pi r^2} \int_{|t|<r} \|\tilde{u}_r\|_t^2 d\lambda_t. \tag{4.60}$$

Hence, in view of (4.55) and (4.58) with $z = z_0$, (4.59) implies

$$1 \geq \exp \frac{1}{\pi r^2} \int_{|t|<r} (2 \log |\tilde{u}_r(t, z_0)/(dz_1 \wedge dz_2 \wedge \ldots \wedge dz_n)| - \log k(t, z_0)) d\lambda_t. \tag{4.61}$$

by the convexity of the function $y = e^x$. But $\log |\tilde{u}_r(t, z_0)/(dz_1 \wedge dz_2 \wedge \ldots \wedge dz_n)|$ is harmonic, so that, in view of (4.56) and the first half of (4.57), the desired sub-mean-value property

$$\log k(0, z_0) \leq \frac{1}{\pi r^2} \int_{|t|<r} \log k(t, z_0) d\lambda_t \tag{4.62}$$

follows from (4.60). □

In 2014, Cao [CJ-2] generalized the above argument to give an alternate proof of Theorem 4.26 for the case $m = 1$, and Berndtsson and Lempert [Brd-L] found a remarkable connection between the plurisubharmonicity of $\log k_D$ and Theorem 3.3. To illustrate the idea of [Brd-L], let us give an alternate proof of Theorem 3.2.

Proof of Theorem 3.2. It suffices to show that $\pi k_{\mathbb{D}}(0) \geq 1$, or equivalently that $\log(\pi k_{\mathbb{D}}(0)) \geq 0$. For that, we put

$$\mathbb{D}_\zeta = \{z; |z| < |\zeta|\} \quad \text{for} \quad |\zeta| < 1.$$

Then, by the Maitani-Yamaguchi theorem (Theorem 4.22), $\log k_{\mathbb{D}_\zeta}(0)$ is subharmonic in $\log \zeta$. Since $\log k_{\mathbb{D}_\zeta}(0)$ depends only on $\log |\zeta|$, it is convex in $\log |\zeta|$. Therefore, the function

$$-\log \pi k_{\mathbb{D}}(0) + \log \pi k_{\mathbb{D}_\zeta}(0) + \log |\zeta|^2$$

turns out to be increasing in $\log |\zeta|$, since it is convex and bounded. Hence

$$\lim_{\zeta \to 0} -\log \pi k_{\mathbb{D}}(0) + \log \pi k_{\mathbb{D}_\zeta}(0) + \log |\zeta|^2 \leq \lim_{\zeta \to 1} -\log \pi k_{\mathbb{D}}(0) + \log \pi k_{\mathbb{D}_\zeta}(0)$$

$$+ \log |\zeta|^2 = 0.$$

Admitting that Theorem 3.2 is infinitesimally true, i.e. that $\lim_{\zeta \to 0} \pi k_{\mathbb{D}_\zeta}(0)|\zeta|^2 = 1$, we obtain the desired conclusion. □

Theorems 3.3 and 3.4 can be deduced from Theorem 4.26 by generalizing the above argument. See [Brd-L] for the detail (and also [Oh-35] for an expository account).

Chapter 5
L^2 Approaches to Holomorphic Foliations

Abstract Some results on the L^2 $\bar{\partial}$-cohomology groups are applied to holomorphic foliations. A basic general result is a nonexistence theorem for the foliation on n-dimensional compact Kähler manifolds whose stable set is a real hypersurface with $(n-2)$-convex and pseudoconvex complement. For the special cases such as $\mathbb{CP}^n$, complex tori and Hopf surfaces, nonexistence, reduction and classification theorems will be proved. Closely related materials have been already discussed in Sect. 2.4, e.g. Theorem 2.79.

5.1 Holomorphic Foliation

Geometric structures of holomorphic foliations on complex manifolds are reflected in the curvature properties of the normal bundle, as in the case of submanifolds. Some results on the curvature of holomorphic foliations of codimension one are discussed. Ghys's turbulent foliations on complex tori and Nemirovski's example on torus bundles are analyzed in terms of meromorphic connections.

5.1.1 Foliation and Its Normal Bundle

By definition, a foliation on a differentiable manifold M is a (possibly disconnected) manifold F with a bijective embedding $\iota : F \to M$ such that $T_F = \iota^* T_{\tilde{F}}$ holds for some differentiable subbundle $T_{\tilde{F}}$ of T. F is called a **foliation** of class C^r if $T_{\tilde{F}}$ is of class C^r. Connected components of F are called the **leaves** of F. If M is a complex manifold and $T_{\tilde{F}}$ is a holomorphic subbundle of T_M, F will be called a **holomorphic foliation** on M. For simplicity we shall not distinguish $T_{\tilde{F}}$ from T_F.

Let F be a holomorphic foliation of codimension r on a complex manifold M of dimension n. Then, for any point $x \in M$, one can find a neighborhood $U \ni x$ and holomorphic 1-forms $\omega_1, \dots, \omega_r$ on U which are pointwise linearly independent and annihilated by T_F. $\omega_j (1 \le j \le r)$ locally generates a subsheaf of $\mathcal{O}((T_M^{1,0})^*)$ which we shall denote by Ω_F for simplicity. The collection of local generators ω_j of Ω_F defines by $\omega_1 \wedge \cdots \wedge \omega_r$ a global holomorphic section of the projectivization $(\wedge^r (T_M^{1,0})^* - \{0\})/\mathbb{C}^*$ of $\wedge^r (T_M^{1,0})^*$. The system of locally defined r-forms $\omega_1 \wedge \cdots \wedge$

© Springer Japan 2015

T. Ohsawa, L^2 *Approaches in Several Complex Variables*, Springer Monographs in Mathematics, DOI 10.1007/978-4-431-55747-0_5

ω_r is naturally identified with a globally defined nowhere–vanishing r-form with values in the bundle $\det N_F$ $(=\wedge^r N_F)$, where N_F denotes the normal bundle of F (in M). Recall that N_F is defined as the quotient of $T_M|_F (\cong T_M^{1,0}|_F)$ by $T_F (\cong T_F^{1,0})$. (N_F will be also denoted by $N_F^{1,0}$.) Clearly, $\mathcal{O}(N_F^*) \cong \Omega_F$. It is easy to see that one may take as ω_j 1-forms of the form df_j for some $f_j \in \mathcal{O}(U)$ by shrinking U if necessary. Hence F is locally a collection of the level sets of $\mathbb{C}^r$-valued holomorphic functions.

Concerning the normal bundle of compact leaves, the following is basic.

Proposition 5.1. *Let F be a holomorphic foliation of codimension $r < n$ on a complex manifold M, and let L be a compact leaf of F. If L admits a Kähler metric, then $c_1(N_L) = 0$.*

Proof. Since L is locally the level of a $\mathbb{C}^r$-valued holomorphic function, the transition functions of N_L associated to them are locally constant. Therefore, by the well–known $\partial\bar{\partial}$-lemma on compact Kähler manifolds (cf. [W, Chapter 6, Proposition 2.2]), $c_1(N_L) = 0$.   □

Example 5.1. Let N be a compact Kähler manifold and let $A \to N$ be a holomorphic affine line bundle, i.e. a fiber bundle whose fibers are $\mathbb{C}$ and transition functions are of the form $\zeta_\alpha = e^{i\theta_{\alpha\beta}}\zeta_\beta + a_{\alpha\beta}$ ($\theta_{\alpha\beta} \in \mathbb{R}$, $a_{\alpha\beta} \in \mathbb{C}$) with respect to an open covering $\{U_\alpha\}$ of N and the fiber coordinates ζ_α over U_α. Then A admits a holomorphic foliation of codimension one whose leaves are locally the level sets of ζ_α. By compactifying A by adding the section at infinity, one has a compact Kähler manifold with a foliation $\zeta_\alpha = const \in \mathbb{C} \cup \{\infty\}$. The section at infinity is then a compact leaf whose normal bundle is topologically trivial.

As for N_F, let us mention some curvature properties.

Proposition 5.2. *Let M be a compact complex manifold of dimension n and let F be a holomorphic foliation of codimension $r < n$ on M. Then $\det N_F$ is not negative.*

Proof. If $\det N_F$ were negative, $H^{r,0}(M, \det N_F) = 0$ would hold by the Akizuki–Nakano vanishing theorem (cf. Theorem 2.12). However, as we have seen above, $H^{r,0}(M, \det N_F)$ contains a nonzero element, which is a contradiction.   □

Similarly, one has the following.

Proposition 5.3. *Let (M, F) be as above. If $r = 1$ or $n - 1$, and $T_M^{1,0}$ is trivial, then $\det N_F$ is not positive.*

Proof. If $\det N_F$ were positive, no element of $H^{0,0}(M, \oplus^n \det N_F)$ would be nowhere zero, because positive dimensional analytic sets must intersect with the zeros of holomorphic sections of positive line bundles on compact complex manifolds.   □

Corollary 5.1. *The normal bundle of a holomorphic foliation of codimension one on a complex torus is not positive.*

The reader may suspect that N_F is more or less flat. However, the following phenomenon must not be neglected.

Theorem 5.1. *There exist a two–dimensional compact complex manifold M and a holomorphic foliation F of codimension one on M such that N_F has a fiber metric whose curvature form is positive along F.*

Proof. Let R be a compact Riemann surface of genus ≥ 2, let $\mathbb{D}$ be the open unit disc and let Γ be a discrete subgroup of $Aut\mathbb{D}$ such that $\mathbb{D}/\Gamma \cong R$. Since $Aut\mathbb{D} \subset Aut\hat{\mathbb{C}}$, $\hat{\mathbb{C}}$ being the Riemann sphere, Γ acts also on $\mathbb{D} \times \hat{\mathbb{C}}$ by

$$(z, w) \to (\gamma(z), \gamma(w)) \quad (\gamma \in \Gamma).$$

Let $M = (\mathbb{D} \times \hat{\mathbb{C}})/\Gamma$, let $\pi : \mathbb{D} \times \hat{\mathbb{C}} \to M$ be the natural projection, and let F be the collection of the images of $\mathbb{D} \times \{w\}$ in M by π. Then M is a compact complex manifold of dimension 2 and F is a holomorphic foliation of codimension one on M. To define a fiber metric of N_F, first note that the Bergman metric $(1-|w|^2)^{-2}dw \otimes d\bar{w}$ on $\mathbb{D} \cup (\hat{\mathbb{C}} - \bar{\mathbb{D}})$ is a fiber metric of $N_{F|M-\pi(\mathbb{D}\times\partial\mathbb{D})}$, because of its invariance under Γ. Hence, by multiplying $(1 - |w|^2)^{-2}dw \otimes d\bar{w}$ by a C^∞ function ρ defined by

$$\rho(z, w) = \begin{cases} \left(1 - \left|\frac{z-w}{\bar{w}z-1}\right|^2\right)^2 & \text{if } z, w \in \mathbb{D} \\ \left(1 - \left|\frac{\bar{w}z-1}{z-w}\right|^2\right)^2 & \text{if } z \in \mathbb{D}, \ w \in \hat{\mathbb{C}} - \mathbb{D}, \end{cases}$$

one has a C^∞ fiber metric of N_F, say b. The curvature form of b is positive along the leaves of F since it is twice the Bergman metric along them.   □

Conjecture 5.1. Let M be a compact complex manifold of dimension ≥ 3 and let F be a holomorphic foliation on M of codimension one. Then N_F does not admit a fiber metric whose curvature form is positive along F.

The reader will find several pieces of supporting evidence for it in subsequent sections.

A holomorphic foliation F on a dense open subset U of M is called a **singular holomorphic foliation** on M if the subsheaf of $\mathcal{O}((T_M^{1,0})^*)$ generated by holomorphic 1-forms annihilated by $T_F^{1,0}$, to be called the **defining sheaf** of F, is locally finitely generated over the structure sheaf of M. For instance, the holomorphic foliation on $\mathbb{C}^2 - \{0\}$ whose tangent bundle is $\text{Ker}(wdz - zdw)$ is a singular holomorphic foliation on $\mathbb{C}^2$. $Sing(F)$ will denote the set of points for which F is not extendible to a holomorphic foliation on their neighborhoods. $Sing(F)$ is called the **singular set** of F. In contrast to the case of holomorphic foliation, not every singular holomorphic foliation is locally expressed as the level set of a vector–valued holomorphic function. Concerning singular holomorphic foliations of codimension one, it is easy to see that the defining sheaf is invertible, so that the normal bundle N_F is well defined also for a singular holomorphic foliation F. A long–standing open question is whether there exists a singular holomorphic foliation F on $\mathbb{CP}^2$ with a leaf which does not accumulate to any point in $Sing(F)$ (cf. [C-LN-S] and [C]).

A singular holomorphic foliation F on M is said to be a **foliation by rational curves** if for every $x \in M$ there exists a rational curve (a complex space bimeromorphically equivalent to $\hat{\mathbb{C}}$) through x and tangent to F. The following was obtained by M. Brunella [Br-1] in the context of bimeromorphic classification theory of algebraic varieties. (See also [Br-4].)

Theorem 5.2. *Let F be a singular holomorphic foliation of dimension one on a compact Kähler manifold M. Suppose that F is not a foliation by rational curves. Then its canonical bundle $\mathbb{K}_F$ is pseudoeffective (i.e. $\mathbb{K}_F$ admits a singular fiber metric whose curvature current is semipositive.).*

Definition 5.1. A closed set $S \subset M$ is called a **stable set** of a singular foliation F if S is the closure (in M) of the union of some leaves.

Minimal stable sets are particularly of interest. Given a singular holomorphic foliation F of codimension one on a compact complex manifold, the complement of a stable set of F is locally pseudoconvex. Hence, minimal stable sets can arise as the boundary of a locally pseudoconvex domain. Therefore, the L^2 method is naturally expected to be useful to study such foliations.

In the general theory of several complex variables, holomorphic foliations of codimension one first arose in a paper of Grauert [Gra-6], where a locally pseudoconvex domain without nonconstant holomorphic functions was presented as a counterexample to a Levi problem on complex manifolds. In the next subsection, we shall collect some examples of holomorphic foliations of codimension one arising as variants of Grauert's example.

5.1.2   Holomorphic Foliations of Codimension One

Let M be a compact complex manifold of dimension n. A complex manifold of dimension $n + 1$ with a holomorphic foliation of codimension one arises as a relatively compact locally pseudoconvex domain in a differentiable disc bundle over M as follows. Let $\{U_\alpha\}$ be a locally finite open covering of M by open sets and let $\Phi = \{\phi_{\alpha\beta}\}$ be a system of injective holomorphic maps from $\mathbb{D}$ to $\mathbb{C}$ fixing the origin, such that $\phi_{\alpha\beta} \circ \phi_{\beta\gamma} = \phi_{\alpha\gamma}$ holds on a neighborhood of 0 as long as $U_\alpha \cap U_\beta \cap U_\gamma \neq \phi$. Then, by gluing a neighborhood of $U_\alpha \times \{0\}$ in $U_\alpha \times \mathbb{D}$ and that of $U_\beta \times \{0\}$ in $U_\beta \times \mathbb{D}$ by

$$(z, \zeta_\alpha) \sim (z, \zeta_\beta) \iff \zeta_\alpha = \phi_{\alpha\beta}(\zeta_\beta)$$

for $z \in U_\alpha \cap U_\beta$, one obtains a complex manifold, say Ω, containing M as a closed submanifold $\zeta_\alpha = 0$. On Ω one has a holomorphic foliation, say F_Φ locally defined by the level sets of the coordinates on $\mathbb{D}$. If M is a compact Riemann surface of genus ≥ 1, some of such F_Φ can contain infinitely many compact leaves which are mutually homotopically nonequivalent (cf. [U-2]). If $\phi_{\alpha\beta}$ are all rotations around the origin, Ω is a tubular neighborhood of the zero section of the holomorphic line bundle N_M over M. If M is the Riemann sphere $\hat{\mathbb{C}}$, it is easy to see that F_Φ are all equivalent to the fibers of the projection $\pi : \hat{\mathbb{C}} \times \mathbb{D} \to \mathbb{D}$ on a neighborhood of M, so that they are holomorphically convex, but they need not be so if the genus of M is ≥ 1. Indeed, if the rotations are given in such a way that the tensor powers N_M^k are not trivial for any $k \in \mathbb{N}$, then Ω is never holomorphically convex, since the leaves of F_Φ are then dense in the level sets of $|\zeta_\alpha|$ by Kronecker's theorem (or by Dirichlet's pigeon hole principle). There exist holomorphic foliations of a similar kind on complex tori, i.e. holomorphic foliations induced from mutually parallel affine subspaces in $\mathbb{C}^n$. Such foliations will be called **linear foliations**. Leaves are dense in most cases, but there exist cases where the foliation admits a real hypersurface as a stable set. If a leaf is dense in such a hypersurface X, then the complement of X is not holomorphically convex, because it is the union of parallel translates of X. Grauert's observation in [Gra-6] is essentially up to this point.

Let $\mathscr{D} \to M$ be a holomorphic disc bundle. Then, as a domain in the associated $\hat{\mathbb{C}}$-bundle, (the total space of) $\mathscr{D}$ is locally pseudoconvex and bounded by a real-analytic hypersurface which is a stable set of the foliation locally consisting of the constant sections of $\mathscr{D} \to M$, which shall be denoted by $F_\mathscr{D}$. It was proved in [D-Oh-2] that $\mathscr{D}$ is pseudoconvex whenever M is Kählerian (cf. Theorem 2.79 in Chap. 2). We shall see later, independently from [D-Oh-2], that $\mathscr{D}$ is not "too pseudoconvex" if M is Kählerian.

Similarly, let $\mathscr{A} \to M$ be a **holomorphic affine line bundle**, i.e. a holomorphic fiber bundle over M with typical fiber $\mathbb{C}$. Since $Aut\mathbb{C}$ consists of holomorphic affine transformations, $\mathscr{A}$ has its associated line bundle, say $\mathscr{A}_0 \to M$. If M is Kählerian and the first Chern class of $\mathscr{A}_0$ is zero, then $\mathscr{A}$ is equivalent to the bundle whose transition functions are locally constant as maps to $Aut\mathbb{C}$. Therefore, $\mathscr{A}$ is equipped with a holomorphic foliation of codimension one whose leaves are locally the constant sections of $\mathscr{A}$. The following is essentially contained in [U-2].

Proposition 5.4. *If M is a compact Kähler manifold, then the total space of topologically trivial holomorphic affine line bundles over M are pseudoconvex.*

Proof. Since M is Kählerian, the transition functions of $\mathscr{A}$ can be given by

$$\zeta_\alpha = \zeta_\beta e^{i\theta_{\alpha\beta}} + c_{\alpha\beta}, \quad \theta_{\alpha\beta} \in \mathbb{R}, \ c_{\alpha\beta} \in \mathbb{C}.$$

Then, by using the Kählerianity again, one has a system of pluriharmonic functions h_α satisfying $c_{\alpha\beta} = h_\alpha - e^{i\theta_{\alpha\beta}} h_\beta$. Then $|\zeta_\alpha - h_\alpha|^2$ is a well–defined plurisubharmonic exhaustion function on $\mathscr{A}$. In fact, since $\partial\bar{\partial}h_\alpha = 0$ one has

$$i\partial\bar{\partial}|\zeta_\alpha - h_\alpha|^2 =$$

$$i(d\zeta_\alpha \wedge d\bar{\zeta}_\alpha - d\zeta_\alpha \wedge \bar{\partial}\bar{h}_\alpha - \partial h_\alpha \wedge d\bar{\zeta}_\alpha + \partial h_\alpha \wedge \bar{\partial}\bar{h}_\alpha + \bar{\partial}\bar{h}_\alpha \wedge \partial h_\alpha)$$

$$\geq i\partial\bar{h}_\alpha \wedge \bar{\partial}h_\alpha \geq 0.\square$$

As well as in the case of disc bundles, $\mathscr{A}$ are not "too pseudoconvex". This point will also be discussed later.

Disc bundles and affine bundles are variants of tubular neighborhoods of submanifolds, although they are not necessarily their deformations. As a variant of foliations on complex tori consisting of mutually parallel leaves, there exists a distinguished class of holomorphic foliations of codimension one, which will be described below.

Let R be a compact Riemann surface (of any genus), let T be a complex torus and let $\pi : P \to R$ be a principal T-bundle. Let $\mathfrak{g}$ be the Lie algebra pf T. The kernel of the exponential map $\exp : \mathfrak{g} \to T$ will be denoted by $\mathfrak{g}_0$. For simplicity, we put $\exp \zeta = [\zeta]$ and do not distinguish T from $\mathfrak{g}/\mathfrak{g}_0$. By a **meromorphic connection** on $T \to R$, we mean a system of $\mathfrak{g}$-valued meromorphic 1-forms, say $\{\omega_\alpha\}$, associated to an open covering $\{U_\alpha\}$ of R with local trivializations

$$\phi_\alpha : \pi^{-1}(U_\alpha) \to U_\alpha \times T,$$

such that ω_α are defined on U_α and mutually related on $U_\alpha \cap U_\beta$ by

$$\omega_\alpha - \omega_\beta = dc_{\alpha\beta}.$$

Here $c_{\alpha\beta}$ are defined by

$$\phi_\alpha \circ \phi_\beta^{-1}(z, [\zeta]) = (z, [\zeta + c_{\alpha\beta}(z)]).$$

Existence of nontrivial meromorphic connections is a consequence of the classical theory of Riemann surfaces, or by Kodaira's vanishing theorem more directly speaking. Since the difference of adjacent ω_α's are d-exact, the parallel transports of the points in P along the paths in $R \setminus \{$poles of $\omega_\alpha\}$ are well defined, depending only on the homotopy class of the paths. Let S be the set of poles of ω_α. By this parallel transport, any linear foliation on a fiber outside $\pi^{-1}(S)$ yields a foliation on $P \setminus \pi^{-1}(S)$. If its codimension is one, by adding the fibers of π over S, one has an extension of the foliation to that on P. A holomorphic foliation on P arising by such a construction will be called a **turbulent foliation**.

Theorem 5.3 (cf. [Gh]). *Any holomorhic foliation of codimension one on a complex torus is either linear or turbulent.*

Proof. Let F be a holomorphic foliation of codimension one on a complex torus T. Since the assertion is trivially true if $\dim T = 1$, we may assume that $\dim T = n \geq 2$. Then F yields a holomorphic map from T to $\mathbb{CP}^{n-1}$ by associating $T_{F,x}$ to $x \in T$, since the tangent bundle of T is trivial. Let L be the connected component of any smooth fiber of this map. Then L must be a complex torus, since its normal bundle is trivial and hence so is T_L. Clearly, F is linear if $L = T$. That F is turbulent otherwise can be seen by induction by considering the factor space of T by L. $\square$

As for the results on singular holomorphic foliations on T, see [Br-3] and [C-LN].

As we have seen above, some disc–bundles and affine line bundles are naturally equipped with holomorphic foliations of codimension one which are extendible to the associated $\hat{\mathbb{C}}$-bundles. They admit stable real hypersurfaces and/or complex submanifolds which are minimally stable and with pseudoconvex complements. In the turbulent foliations, the preimages of the poles of the meromorphic connection are minimal stable sets. In [Nm], S. Nemirovski discovered that some turbulent foliations can contain real hypersurfaces as stable sets which are not minimal. Let us describe his construction below.

Let N be a compact complex manifold of dimension $m \geq 1$ and let $p : E \to N$ be a holomorphic line bundle. Let s be a meromorphic section of E whose zeros and poles are all of order one along a nonempty smooth submanifold say B of N. Then we put

$$S = \{c \cdot s(x); x \notin B \ \textit{and} \ c > 0\}.$$

Taking the quotient of S by the action of the infinite cyclic group $\mathbb{Z}$ on $E^* = E \setminus \{\text{zero section}\}$ by fiberwise multiplication by 2^a for $a \in \mathbb{Z}$, we obtain a real hypersurface $S/\mathbb{Z}$ in $E^*/\mathbb{Z}$. Since the order of zeros and poles of s is one, the closure of $S/\mathbb{Z}$ in $E^*/\mathbb{Z}$ becomes a smooth real hypersurface. The union of (the images of) the sections $\zeta \cdot s$ ($\zeta \in \mathbb{C} \setminus \{0\}$) over $N \setminus B$ and the preimage of B is a holomorphic foliation of codimension one on E^* which induces a foliation on $E^*/\mathbb{Z}$, and $\overline{S/\mathbb{Z}}$ is a stable set of this foliation.

A remarkable feature of **Nemirovski's hypersurface** $\overline{S/\mathbb{Z}}$ is that its complement is Stein if E is positive and s is everywhere holomorphic. In fact, $N \setminus B$ is then an affine algebraic manifold by Kodaira's embedding theorem, hence it is Stein, and $(E^*/\mathbb{Z}) \setminus (\overline{S/\mathbb{Z}})$ is Stein because it is a holomorphic fiber bundle over a Stein manifold with one-dimensional Stein fibers (cf. [Mk]).

In [Oh-24], a generalization of Nemirovski's construction is given. It turned out that, for a turbulent foliation to admit a stable real hypersurface, the meromorphic connection must satisfy a period condition.

As for the analytic continuation of holomorphic foliations, a positive result was obtained by T. Nishino [Ni-2, Ni-3] when the leaves are compact and of codimension one.

Theorem 5.4 (cf. [Ni-3]). *Let F be a holomorphic foliation of codimension one on a nonempty open subset of a complex manifold M. Suppose that the leaves of F are compact and M is an increasing union of relatively compact locally pseudoconvex domains. Then there exists a holomorphic foliation on M extending F.*

5.2 Applications of the L^2 Method

As was indicated in the preceding section, the method of L^2 estimates can be applied to prove that the Kähler condition imposes certain restrictions on holomorphic foliations. To see this, the structure of L^2 $\bar{\partial}$-chohomology on the locally pseudoconvex domains in Kähler manifolds has to be observed more closely, extending what has been seen in Chap. 2.

5.2.1 Applications to Stable Sets

Let D be a locally pseudoconvex relatively compact domain in a complex manifold. First we shall study the case where ∂D is a smooth real hypersurface of (real) codimension one. Let us first observe that the classical theory of $\bar{\partial}$-cohomology groups on Stein manifolds and compact Kähler manifolds already yields a prototypical result.

Definition 5.2. A closed submanifold of real codimension one in a complex manifold is said to be a **Levi flat hypersurface** if its complement is locally pseudoconvex.

If ∂D is C^2-smooth with a defining function ρ, it is clear that ∂D is Levi flat if and only if $\partial\bar{\partial}\rho|_{(\text{Ker}\partial\rho)\cap(T_{\partial D}\otimes\mathbb{C})} \equiv 0$. By an abuse of language, we shall say that ∂D is Levi flat at $x \in \partial D$ if $\partial\bar{\partial}\rho|_{\text{Ker}\partial\rho} = 0$ at x.

Proposition 5.5. *A Levi flat hypersurface of class C^r ($r \geq 2$) in a complex manifold M admits a foliation of class C^r of real codimension one whose leaves are complex submanifolds in M.*

Proof. Let $X \subset M$ be a Levi flat hypersurface. Then the analytic tangent bundle $T_X^{1,0} = T_X^{\mathbb{C}} \cap T_M^{1,0}|_X$ has a property that $T_X^{1,0} \oplus \overline{T_X^{1,0}}$ is involutive, i.e. it is closed under the Lie bracket, since

$$\partial\rho([\xi,\overline{\eta}]) = \partial\rho([\xi,\overline{\eta}]) - \xi\partial\rho(\overline{\eta}) + \overline{\eta}\partial\rho(\xi) = \partial\bar{\partial}\rho(\xi,\overline{\eta}) = 0$$

if $X = \{\rho = 0\}$ ($\rho \in C^r$) and ξ and η are local C^∞ sections of $T_X^{1,0}$. Therefore, by the Frobenius theorem one has the desired foliation.   □

We shall call the foliation F on X satisfying $T_F^{1,0} = T_X^{1,0}$ the **Levi foliation** on X. The Levi foliation on X will be denoted by L_X. Clearly, N_{L_X} is equivalent to $N_X^{1,0} := (T_M^{1,0}|_X)/T_X^{1,0}$, which is defined for any real hypersurface X and called the **analytic normal bundle** of X.

Proposition 5.6. *A real analytic Levi flat hypersurface locally admits a pluriharmonic defining function.*

Proof. Given a real analytic Levi flat hypersurface X in a complex manifold M of dimension n, let $x \in X$, let U be a neighborhood of x in X, and let $f : U \to \mathbb{R}$ be a real analytic function with $df(x) \neq 0$ whose level sets are contained in the levels of L_X. Then, by shrinking U if necessary, one can find a real analytic equivalence from $(0, 1) \times \mathbb{D}^{n-1}$ to U, say $\alpha(t, z)$, which is holomorphic in the variable $z \in \mathbb{D}^{n-1}$. Then the conclusion is obvious because α can be extended to a biholomorphic equivalence between some neighborhoods of $(0, 1) \times \mathbb{D}^{n-1}$ and U, in $\mathbb{C}^n$ and M, respectively.   □

Let us prove a nonexistence result for Levi flat hypersurfaces which are stable sets of holomorphic foliations.

Theorem 5.5. *Let (M, ω) be a compact Kähler manifold of dimension n whose holomorphic bisectional curvature (see Sect. 2.4.4) is positive. If $n \geq 3$, singular holomorphic foliations of codimension one on M do not admit C^∞ hypersurfaces as stable sets.*

Proof. Suppose that there existed a C^∞ Levi flat hypersurface X in M. Then, by taking the double cover of M if necessary, we may assume that $M \setminus X = D_+ \cup D_-$, where $D_\pm$ are mutually disjoint locally pseudoconvex domains. Then, by a theorem of Takeuchi, Elencwajc and Suzuki (cf. Theorems 2.73 and 2.74), the curvature condition on M implies that $D_\pm$ are Stein domains. Therefore $H^{0,2}_0(D_\pm) = 0$ if $n \geq 3$. Now, suppose moreover that X is a stable set of some holomorphic foliation of codimension one, say F, on a neighborhood of X. Then the normal bundle N_F is topologically trivial on a neighborhood of X because so is $N_X \otimes \mathbb{C}$ which is topologically equivalent to $N_F|_X$. The curvature form, say θ, of the fiber metric of N_F induced from that of $T^{1,0}_M$ is positive along F in virtue of the Gauss–Codazzi–Griffiths formula, since the holomorphic bisectional curvature of M is positive by assumption. Since $N_F|_X$ is topologically trivial, there exists a neighborhood $U \supset X$ and a 1-form η satisfying $\theta = d\eta$ on U. Splitting η into the sum of the $(1,0)$-component $\eta^{1,0}$ and the $(0,1)$-component $\eta^{0,1}$, one has $\bar\partial\eta^{0,1} = 0$, because θ is of type $(1,1)$. Since $H^{0,2}_0(D_\pm) = 0$, $\eta^{0,1}$ can be extended from a neighborhood of X to M as a $\bar\partial$-closed $(0,1)$-form, say $\tilde\eta$. Since M is a compact Kähler manifold, the harmonic representative of $\tilde\eta$ is ∂-closed. This means that $\theta = \partial\bar\partial\phi$ holds for some C^∞ function ϕ, which is absurd because of the compactness of X and the maximum principle for the plurisubharmonic functions on F. □

Remark 5.1. The idea of the above proof for the C^∞ case is taken from Siu [Siu-8]. Nonexistence of Levi flat hypersurfaces in $\mathbb{CP}^n$ for $n \geq 3$ was first proved by Lins Neto in [LN] for the real analytic case by a method independently from the L^2 method, and by Siu [Siu-8], Cao and Shaw [C-S] and Brunella [Br-2] for less regular cases. The latter works are based on the L^2 method. It was asked in [C] whether or not $\mathbb{CP}^2$ contains a Levi flat hypersurface. It is known that if it does, then the hypersurface has to satisfy a seemingly very restrictive curvature condition (cf. [A-B]).

Let D be a locally pseudoconvex relatively compact domain in a Kähler manifold (M, ω) of dimension n. Although it is not known whether or not D carries a plurisubharmonic exhaustion function, definite results still hold for the $\bar\partial$-cohomology of such D. Some of them can be applied to study holomorphic foliations.

Let ρ be a defining function of D, i.e. ρ is a C^∞ function defined on a neighborhood U of $\overline{D}$ such that $D = \{x \in U; \rho(x) < 0\}$ and $d\rho$ vanishes nowhere on ∂D. We shall analyze the $\bar\partial$-cohomology of D by the L^2 method under some restrictions on the eigenvalues of the Levi form $\partial\bar\partial(-\log(-\rho))$ of $-\log(-\rho)$. These conditions implicitly appeared above when ∂D is a stable set of a holomorphic foliation on complex manifolds with certain curvature properties such as $\mathbb{CP}^n$. Let $\lambda_1 \geq \lambda_2 \geq \ldots \geq \lambda_n$ be the eigenvalues of $i\partial\bar\partial(-\log(-\rho))$ with respect to ω. Since $\partial\bar\partial(-\log(-\rho)) = -\partial\bar\partial\rho/\rho + \partial\rho \wedge \bar\partial\rho/\rho^2$,

$$\liminf_{x\to\partial D} \rho(x)^2\lambda_1 > 0. \tag{5.1}$$

Proposition 5.7. *Suppose that*

$$\liminf_{x\to\partial D} \rho(x)^2\lambda_p > 0 \ for \ 2 \leq p \leq k, \tag{5.2}$$

and

$$\liminf_{x\to\partial D} \rho(x)^2\lambda_p \geq 0 \ for \ k+1 \leq p \leq n \tag{5.3}$$

hold for some k. Then, for any Nakano semipositive vector bundle E over M,
$H^{n,p}(D,E) = 0$ *for* $p \geq n-k+1$ *and* $H_0^{0,p}(D,E^*) = 0$ *for* $p \leq k-1$.

Proof. By assumption, $\omega + \epsilon i\partial\bar\partial(-\log(-\rho))$ is a complete Kähler metric on D for sufficiently small $\epsilon > 0$. By (5.1), (5.2) and (5.3), $\lambda_k+\lambda_{k+1}+\cdots+\lambda_n > 0$ outside a compact subset of D. Therefore, one can find $c \in \mathbb{R}$ and a C^∞ function $\mu : \mathbb{R} \to \mathbb{R}$ satisfying

$$\mu|_{(-\infty,c]} \equiv 0, \quad \mu'|_{(c,\infty)} > 0, \quad \mu''|_{(c,\infty)} > 0$$

such that the sum of $n-k+1$ eigenvalues of $i\partial\bar\partial\mu(-\log(-\rho))$ are nonnegative everywhere and positive on $\{x \in D; \rho(x) > -e^{-c}\}$. Hence, as in the proof of Theorem 2.42 obtained from Corollaries 2.10 and 2.11, recalling (2.17) one has similarly $H^{n,p}(D,E) = 0$ for any $p \geq n-k+1$. That $H_0^{0,p}(D,E^*) = 0$ for $p \leq k-1$ follows from this by Serre's duality theorem. □

Theorem 5.6 (cf. [Oh-29]). *Let (M,ω) be a compact Kähler manifold of dimension n and let $X \subset M$ be a real analytic Levi flat hypersurface. Then $N_X^{1,0}$ does not admit a fiber metric whose curvature form is semipositive of rank ≥ 2 everywhere along L_X.*

Proof. Let us take a locally finite open covering $\{U_j\}$ of X and real analytic functions $f_j : U_j \to \mathbb{R}$ such that T_{L_X} is locally equal to Ker df_j. We may assume that $f_j = \text{Re } h_j$ for some holomorphic function h_j on a neighborhood of U_j. Note that Im h_j is then a local defining function of X. Suppose that $N_X^{1,0}$ had admitted a fiber metric with semipositive curvature of rank ≥ 2. Then one would have a system of C^∞ positive functions a_j on U_j such that $a_k = |\frac{df_j}{df_k}|^2 a_j$ holds on $U_j \cap U_k$, by taking a refinement of $\{U_j\}$ if necessary, such that $-i\partial\bar\partial \log a_j$ is a positive (1,1)-form on $L_X \cap U_j$ for each j. Since $a_k|\text{Im } h_k|^2 - a_j|\text{Im } h_j|^2$ vanishes along $U_j \cap U_k$ with order at least 3, one can find defining functions of the components of $M \setminus X$ satisfying the conditions (5.1), (5.2) and (5.3) for $k = 3$. On the other hand, it follows also from the real analyticity of X that X is a stable set of a holomorphic foliation of codimension one on a neighborhood of X. Hence, similarly to the proof of Theorem 5.5, we obtain the conclusion. □

In view of Theorem 5.1, the condition on the rank of the curvature form of $N_X^{1,0}$ is optimal. In particular, the boundary of holomorphic disc bundles over compact Kähler manifolds cannot carry positive analytic normal bundles. Thus, Theorem 5.6 may well be regarded as a supporting evidence for Conjecture 5.1. Now, turning our attention from the geometry of Levi flat hypersurfaces to that of the domains they bound, it is natural to ask for their q-convexity properties. The following answer may be regarded as a pseudoconvex counterpart of Theorem 5.6.

Theorem 5.7 (cf. [Oh-22]). *Let (M, ω) be a compact Kähler manifold and let $X \subset M$ be a real analytic Levi flat hypersurface. Then there exist no plurisubharmonic exhaustion functions on $M \setminus X$ whose Levi form has everywhere at least three positive eigenvalues outside a compact subset of $M \setminus X$.*

Proof. By assumption, one can find a neighborhood $U \supset X$ and a holomorphic foliation F on U of codimension one extending L_X. Then, let $\{df_\alpha\}$ be a system of holomorphic 1-forms associated to an open covering $\{U_\alpha\}$ of U such that $T_F^{1,0}$ is locally equal to $\mathrm{Ker}\, df_\alpha$. Let $df_\alpha = e_{\alpha\beta} df_\beta$ hold on $U_\alpha \cap U_\beta$. $\{e_{\alpha\beta}\}$ is a system of transition functions of N_F, and $\{df_\alpha\}$ is naturally identified with an N_F-valued 1-form. By shrinking U if necessary, we may assume that N_F is topologically trivial, so that one may assume that $e_{\alpha\beta} = e^{u_{\alpha\beta}}$ for some additive cocycle $\{u_{\alpha\beta}\}$ of holomorphic functions. Since $H_0^{0,2}(M \setminus X) = 0$ by assumption, the proof being similar to that in Proposition 5.7, N_F is extendible as a topologically trivial holomorphic line bundle over M, say $\tilde{N}_F$. Now, concerning the $\tilde{N}_F$-valued $\bar\partial$-cohomology on $M \setminus X$, the pseudoconvexity assumption on $M \setminus X$ allows us to extend Theorem 2.49 to conclude that the map $\wedge^{n-2}\omega : H_0^{1,1}(M \setminus X, (\tilde{N}_F)^*) \to H^{n-1,n-1}(M \setminus X, (\tilde{N}_F)^*)$ is an isomorphism, so that the natural homomorphism

$$H_0^{1,1}(M \setminus X, \tilde{N}_F) \to H^{1,1}(M \setminus X, \tilde{N}_F)$$

is injective. Therefore, $\{\omega_\alpha\}$ is extendible as an $\tilde{N}_F$-valued d-closed holomorphic 1-form on M. Therefore, we may assume in advance that $\{df_\alpha\}$ are related by

$$df_\alpha = e^{i\theta_{\alpha\beta}} df_\beta, \quad \theta_{\alpha\beta} \in \mathbb{R}.$$

Hence one can measure the distance $d(x)$ from a point $x \in U$ to X with respect to $df_\alpha \otimes \overline{df_\alpha}$. More explicitly, one may put $d(x) = \inf_{\alpha,c} |f_\alpha(x) + c|$. Here α is chosen so that $x \in U_\alpha$ and, for each α, c runs through the complex numbers satisfying $\inf_{y \in X \cap U_\alpha} |f_\alpha(y) + c| = 0$. (Note that $f_\alpha = e^{i\theta_{\alpha\beta}} f_\beta + \eta_{\alpha\beta}$ on $U_\alpha \cap U_\beta$ for some $\eta_{\alpha\beta} \in \mathbb{C}$, where $U_\alpha \cap U_\beta$ are implicitly chosen to be connected.) By shrinking U if necessary, we may assume that d is constant on $f_\alpha^{-1}(\zeta)$ for any $\zeta \in f_\alpha(U_\alpha)$, so that the levels of d are compact and foliated by complex submanifolds of M of codimension one near X, which obviously contradicts the maximum principle for the assumed exhaustion function on $M \setminus X$. $\qquad\square$

Remark 5.2. From the last part of the proof, the reader may well have an impression that the assumption on the number of positive eigenvalues of the exhaustion function might be superfluous. However, the number 3 is optimal, as the example in Theorem 5.1 shows.

There is a straightforward extension of Theorem 5.7 for the stable sets of certain singular holomorphic foliations. The following is essentially a repetition of what Theorem 5.7 says.

Theorem 5.8. *Let (M, ω) be a compact Kähler manifold of dimension n and let $X \subset M$ be a closed set. Suppose that there exist a neighborhood $U \supset X$ and a singular holomorphic foliation F of codimension one on U having X as a stable set, such that the defining sheaf of F is locally generated by a d-closed form. Then, $M \setminus X$ does not admit a plurisubharmonic exhaustion function whose Levi form has at most $n - 3$ nonpositive eigenvalues everywhere on $U \setminus X$.*

When X is a divisor, a somewhat stronger theorem holds. For the proof, which reduces the situation to that of the above theorem, one needs a property of a subsheaf of the germs of holomorphic 1-forms consisting of d-closed ones. By a standard argument of algebraic geometry, it is shown that X would extend to a singular foliation on M such that X does not intersect any other leaf, if $M \setminus X$ is too pseudoconvex. Similarly to Theorems 5.7 and 5.8, one has the following.

Theorem 5.9 (cf. [Oh-23]). *Let M be a compact Kähler manifold and let $D \subset M$ be a domain. Suppose that $B := M \setminus D$ is a complex analytic set of codimension one such that there exists an effective divisor A with support B for which the line bundle $[A]|_B$ is topologically trivial. Then D does not admit a plurisubharmonic exhaustion function whose Levi form has at most $n - 3$ nonpositive eigenvalues everywhere outside a compact subset of D.*

For the detail of the proof, see [Oh-23]. Theorem 5.9 is also a supporting evidence of Conjecture 5.1.

In the above study of stable sets of holomorphic foliations, the L^2 method played a role in extending the $\bar{\partial}$-cohomology classes and $\bar{\partial}$-closed forms. Therefore, in order to put the argument into a wider scope, we shall make a digression in the next subsection to prove several Hartogs–type extension results on complex manifolds by the L^2 method. The author is inclined to believe that these general results can be applied not only to foliations but also to other questions in several complex variables, dynamical systems for instance.

5.2.2   Hartogs–Type Extensions by L^2 Method

In this subsection, we shall restrict ourselves to the study of extension phenomena for holomorphic functions on complex manifolds. The following classical theorem attributed to Bochner and Hartogs is our prototype.

Theorem 5.10. *Let M be a Stein manifold of dimension ≥ 2, let $K \subset M$ be a compact set with connected complement. Then every holomorphic function on $M \setminus K$ has a holomorphic extension to M.*

Proof. Since M is Stein and $\dim M \geq 2$, $H_0^{0,1}(M) = 0$ by Theorem 2.22. $\qquad\square$

By this proof it is clear that Theorem 5.10 is also true for any $(n-1)$-complete n-dimensional complex manifold M (cf. Corollary 2.14). We note that the result was extended to $(n-1)$-complete spaces by a different method (cf. [M-P]).

Now, let D be a relatively compact locally pseudoconvex domain in a complex manifold M of dimension n. We ask for a condition on ∂D for a similar extendibility result to hold on D.

Theorem 5.11 (cf. [Oh-26]). *In the above situation, assume that M admits a Kähler metric. Then $H_0^{0,1}(D) = 0$ in the following cases.*

Case I. ∂D *is a C^2-smooth real hypersurface and not everywhere Levi flat.*

Case II. *There exists an effective divisor E on M with $|E| = \partial D$ such that the line bundle $[E]$ admits a fiber metric whose curvature form restricted to the Zariski tangent spaces of ∂D is semipositive everywhere but not identically 0.*

For the proof of Case I, let us prepare the following elementary lemma.

Lemma 5.1. *Let ρ be a real-valued C^2-function on the closed unit disc $\{z \in \mathbb{C}; |z| \leq 1\}$ such that $\rho(0) = 0$ and $d\rho$ vanishes nowhere, let $U = \{z; |z| < 1$ and $\rho(z) > 0\}$, and let f be a holomorphic function on U. Suppose that*

$$\int_U |f(z)|^2 / \rho(z) d\lambda < \infty.$$

Here $d\lambda$ denotes the Lebesgue measure. Then there exists $r > 0$ such that $f(z) = 0$ on $U \cap \{|z| < r\}$.

Proof. By a coordinate change, we may assume in advance that U is connected and

$$\{z \in \mathbb{D}; y < -2x^2\} \subset U \subset \{y < -x^2\} \quad (x = \operatorname{Re} z \text{ and } y = \operatorname{Im} z).$$

For $0 < a < 1$ and $A > 0$ we put

$$f_{a,A}(z) = e^{-Ai(z+ia)/(1-iaz)} f(z).$$

Note that

$$f_{a,A}(-ia) = f(-ia) \quad \text{if} \quad -ia \in U. \tag{5.4}$$

Since

$$-i\frac{z+ia}{1-iaz} = \frac{-i(x+iy+ia)}{1-ia(x+iy)} = \frac{(y+a)-ix}{1-iax+ay},$$

one has

$$\text{Re}\left(-i\frac{z+ia}{1-iaz}\right) = \frac{a+(a^2+1)y+a(x^2+y^2)}{(1+ay)^2+a^2x^2}.$$

Therefore one can find $\epsilon > 0$ and $a_0 > 0$ so that $-ia \in U$ and

$$\text{Re}\left(-i\frac{z+ia}{1-iaz}\right) < -\frac{\epsilon}{2} \qquad\qquad (5.5)$$

holds if $0 < a < a_0$, $|x| < 1$ and $y < -\epsilon$.

Since

$$\int_U |f(z)|^2/\rho(z)d\lambda < \infty,$$

one has

$$\liminf_{\delta \to 0} \int_{\rho=\delta} |f(z)||dz| = 0.$$

Hence, given $0 < a < a_0$ and $N > 0$, one can choose a $\delta > 0$ in such a way that $\rho(-ia) > \delta$ and

$$\int_{\rho=\delta} |f(z)||dz| < \frac{1}{N}.$$

Thus, for any $a \in (0, a_0)$, one can find sequences $\delta_\mu \to 0$ and $A_\mu \to \infty$ such that

$$\lim_{\mu\to\infty} \int_{\rho=\delta_\mu} |f_{a,A_\mu}(z)||dz| = 0. \qquad\qquad (5.6)$$

On the other hand, since $A_\mu \to \infty$, one has

$$\liminf_{\mu\to\infty} \int_{\{|z|=1-\delta_\mu\}\cap U} |f_{a,A_\mu}(z)||dz| = 0 \qquad\qquad (5.7)$$

by (5.5). Consequently one has $f(-ia) = \lim_{\mu\to\infty} f_{a,A_\mu}(-ia) = 0$ if $0 < a < a_0$, so that $f \equiv 0$ by the theorem of identity. $\square$

Remark 5.3. By the boundary regularity in Riemann's mapping theorem, Lemma 5.1 is also an immediate consequence of the Poisson-Jensen formula.

Proof of Case I. Let δ be the distance to the boundary of ∂D with respect to a Kähler metric, say ω, on M, and let ρ be a negative C^∞ function on D such that $\delta + \rho$

vanishes along ∂D at least to the second order. For sufficiently large A, we put $\omega_A = A\omega - i\partial\bar{\partial}(1/\log(-\rho))$ so that ω_A is a complete Kähler metric on D. Such an A exists because

$$\partial\bar{\partial}(1/\log(-\rho)) = \frac{\partial\bar{\partial}\rho}{\rho(\log(-\rho))^2} - \left\{ \frac{1}{\rho^2(\log(-\rho))^2} + \frac{2}{\rho^2(\log(-\rho))^3} \right\} \partial\rho \wedge \bar{\partial}\rho.$$
(5.8)

By the above lemma, it suffices to prove that $H^{0,1}_{(2)}(D) = 0$ with respect to ω_A. Since $1/\log(-\rho)$ is bounded, ω_A is complete, and the sum of n eigenvalues of $i\partial\bar{\partial}(1/\log(-\rho))$ with respect to ω_A is bounded from below by a positive constant near ∂D, we know already that $H^{0,1}_{(2)}(D)$ is Hausdorff (cf. Theorems 2.13 and 2.4). Therefore, in virtue of Aronszajn's unique continuation theorem, it suffices to show that there exist no nonzero L^2 harmonic forms of type $(0,1)$ with respect to ω_A and the fiber metric e^τ of the trivial bundle, for some C^∞ bounded function τ on D. To find such a weight function τ, let us first take a Levi non-flat point $x \in \partial D$ and a compactly supported nonnegative C^∞ function χ on M such that $\chi(x) = 1$ and ∂D is nowhere Levi flat on $\partial D \cap \text{supp } \chi$. Then we put $\tau = \lambda(\epsilon\chi + 1/\log(-\rho))$ for $\epsilon > 0$ and a C^∞ weakly convex increasing function λ such that $\lambda(t) = 0$ for $t < 0$ and $\lambda''(t) > 0$ for $t > 0$. Then, in view of (5.8), it is easy to see that for sufficiently small ϵ, for any choice of $(n-1)$ eigenvalues $\tau_1, \ldots \tau_{n-1}$ of $i\partial\bar{\partial}\tau$ with respect to ω_A, $\sum_{j=1}^{n-1} \tau_j$ is nonnegative everywhere and positive on supp $\chi \cap D$. Therefore, by (2.19) and the basic inequality $\|\bar{\partial}u\|^2 + \|\bar{\partial}^*u\|^2 \geq (i\Lambda\partial\bar{\partial}\tau \wedge u, u)$ for $u \in C^{0,1}_0(D)$ (and by recalling Gaffney's theorem again), we obtain the conclusion. $\square$

The proof of Case II is similar. Since the construction of the metric and the weight function on D is more involved, the reader is referred to [Oh-26] for the detail.

Remark 5.4.   It is likely that the Kähler assumption is superfluous in Theorem 5.11. As supporting evidence, let us mention a result in [D-Oh-1] which asserts that two-dimensional locally pseudoconvex bounded domains with real analytic, Levi non-flat and connected boundary are holomorphically convex.

5.3   Levi Flat Hypersurfaces in Tori and Hopf Surfaces

Although it is still a big open problem whether or not $\mathbb{CP}^2$ contains no Levi flat hypersurfaces, the nonexistence result on $\mathbb{CP}^n$ for $n \geq 3$ due to Lins Neto [LN] was generalized in several ways, as we have seen in Sect. 5.2.1. Therefore it is natural to proceed this way to see what can be said for Levi flat hypersurfaces in other typical complex manifolds. We shall present such results in the case of several complex homogeneous manifolds and their deformations.

5.3.1   Lemmas on Distance Functions

Let Ω be a domain of holomorphy in $\mathbb{C}^n$ with $\partial\Omega \neq \emptyset$ and let $\delta : \Omega \to]0, \infty[$ be the Euclidean distance to $\partial\Omega$. Recall that $-\log \delta$ is plurisubharmonic on Ω. This follows from a more basic fact that, for each $v \in \mathbb{C}^n - \{0\}$, the Euclidean distance from $z \in \Omega$ to $\partial\Omega \cap \{z + \zeta v; \zeta \in \mathbb{C}\}$, say δ_v, has a property that $-\log \delta_v$ is, as a function on Ω with values in $[-\infty, \infty[$, plurisubharmonic. The functions δ and δ_v are naturally generalized on the domains in complex manifolds; δ of course makes sense on Hermitian manifolds and so does δ_v on holomorphically foliated Hermitian manifolds, while v has to be replaced by the foliation. In the latter case, δ_v is generalized as the distance along the leaves with respect to a semipositive $(1,1)$-form inducing a metric on each leaf of the foliation. Let us consider a specific situation where Ω is a pseudoconvex domain in the ball $\mathbb{B}^n = \{z \in \mathbb{C}^n; \|z\| < 1\}$ defined by an inequality $\operatorname{Re} f(z) > 0$, where $f(z)$ is a holomorphic function on $\mathbb{B}^n$ such that $f(z) - z_n = O(2)$ at the origin $z = 0$. Here $O(v)$ is Laudau's symbol for $v \in \mathbb{N}$. Let $\delta_0(z)$ denote the distance from $z = (z', z_n) \in \Omega$ to $\partial\Omega \cap \{(z', \zeta); \zeta \in \mathbb{C}\}$ with respect to the Euclidean metric.

Lemma 5.2 (cf. [Oh-31]).   *There exists $\varepsilon > 0$ such that*

$$\operatorname{rank}(\partial\bar{\partial}(-\log \delta_0)_{|\operatorname{Ker} dz_n})_{|z=(0,t)} = \operatorname{rank}(df_{|\operatorname{Ker} dz_n})_{|z=0}$$

holds if $(0, t) \in \Omega$ and $0 < t < \varepsilon$.

Proof. Clearly, it suffices to prove the assertion for $n = 2$. So we set

$$f(z) = z_2 e^{iL(z)} + c z_1^2 + O(3),$$

where $L(z) = az_1 + bz_2$. Then, for sufficiently small $t > 0$, the Taylor expansion of δ_0 at $(0, t)$ is calculated as

$$\delta_0(z_1, z_2) = t \cos(\operatorname{Re} L(z)) + \operatorname{Re}(z_2 - t + c z_1^2) + O(3)$$

$$= t - t(\operatorname{Re} L(z))^2/2 + \operatorname{Re}(z_2 - t + c z_1^2) + O(3).$$

Hence

$$4t^2(-\partial\bar{\partial}\log \delta_0)_{|z=(0,t)} = (t\partial L\bar{\partial}\bar{L} + 2dz_2 d\bar{z}_2)_{|z=(0,t)}.$$

Therefore $a = 0$ if and only if $\operatorname{rank}(\partial\bar{\partial}(-\log \delta_0)_{|\operatorname{Ker} dz_2}) = 0$ at $(0, t)$. $\square$

As an immediate consequence one has the following.

Proposition 5.8.   *The Levi form of the level set of $\delta_0(z)$ at $(0, \zeta) \in \Omega$ is zero for sufficiently small ζ if and only if $\partial^2 f/\partial z_j \partial z_n(0) = 0$ for $j = 1, 2, \ldots, n - 1$.*

Putting Proposition 5.8 in another way, we obtain:

Theorem 5.12. *Let Ω and f be as above. If there exists $\varepsilon > 0$ such that*

$$\operatorname{rank} \partial\bar{\partial}(-\log \delta_0) = 1$$

on $\Omega \cap \{(0, \zeta); 0 < |\zeta| < \varepsilon\}$, then there exists $\varepsilon' > 0$ such that $\partial f = dz_n$ holds at $(0, \zeta)$ if $|\zeta| < \varepsilon'$.

Thus we know how $-\log \delta_0$ detects the tilt of the leaves of $L_{\partial\Omega}$. It is expected that $-\log \delta$ also has such a property. K. Matsumoto [M] made it explicit in the case where $\partial\Omega$ is the graph of a holomorphic function, i.e. when f is of the form $z_n - g(z_1, \ldots, z_{n-1})$. In this situation, let $\Omega = \{f \neq 0\}$, let

$$G = \left(\frac{\partial^2 g}{\partial z_j \partial z_k}(0) \right)_{1 \leq j,k \leq n-1}$$

and let

$$\Phi(\zeta) = \left(\frac{\partial^2(-\log \delta)}{\partial z_j \partial \bar{z}_k}(0, \zeta) \right)_{1 \leq j,k \leq n-1}.$$

Lemma 5.3. *There exists $\varepsilon > 0$ such that*

$$\Phi(\zeta) = \frac{1}{2}\bar{G}G(I - \bar{G}G)^{-1}$$

holds for $0 < |\zeta| < \varepsilon$. Here I denotes the identity matrix.

Proof. Let $w = (w_1, \ldots, w_{n-1})$ and put

$$\alpha(z, w) = \sum_{j=1}^{n-1} |z_j - w_j|^2 + |z_n - g(w)|^2.$$

Then, on a neighborhood of 0, $\delta(z)^2$ is characterized as $\alpha(z, w(z))$, where $w(z)$ is the solution to the functional equation

$$\frac{\partial\alpha}{\partial w_j}(z, w(z)) = 0 \quad \text{and} \quad \frac{\partial\alpha}{\partial \bar{w}_j}(z, w(z)) = 0 \quad (1 \leq j \leq n-1). \tag{5.9}$$

Therefore

$$\frac{\partial\delta^2}{\partial z_j} = \frac{\partial\alpha}{\partial z_j} = \bar{z}_j - \bar{w}_j \quad (1 \leq j \leq n-1)$$

and

$$\frac{\partial^2\delta^2}{\partial z_j \partial z_k} = \delta_{jk} - \frac{\partial \bar{w}_j}{\partial \bar{z}_k} \quad (1 \leq j \leq n-1).$$

By differentiating (5.9) one has

$$\frac{\partial^2 \alpha}{\partial w_j \partial z_k} + \sum_{\ell=1}^{n-1} \left(\frac{\partial^2 \alpha}{\partial w_j \partial w_\ell} \frac{\partial w_\ell}{\partial z_k} + \frac{\partial^2 \alpha}{\partial w_j \partial \bar{w}_\ell} \frac{\partial \bar{w}_\ell}{\partial z_k} \right) = 0$$

and

$$\frac{\partial^2 \alpha}{\partial \bar{w}_j \partial z_k} + \sum_{\ell=1}^{n-1} \left(\frac{\partial^2 \alpha}{\partial \bar{w}_j \partial w_\ell} \frac{\partial w_\ell}{\partial z_k} + \frac{\partial^2 \alpha}{\partial \bar{w}_j \partial \bar{w}_\ell} \frac{\partial \bar{w}_\ell}{\partial z_k} \right) = 0.$$

On the other hand, from the definition of α, one has

$$\frac{\partial^2 \alpha}{\partial w_j \partial z_k} = 0, \qquad \frac{\partial \alpha}{\partial \bar{w}_j \partial z_k} = -\delta_{jk},$$

$$\frac{\partial^2 \alpha}{\partial w_j \partial w_k} = \frac{\partial^2 g}{\partial w_j \partial w_k} (g(w) - z_n)$$

and

$$\frac{\partial^2 \alpha}{\partial w_j \partial \bar{w}_k} = \delta_{jk} + \frac{\partial g}{\partial w_j} \frac{\partial \bar{g}}{\partial \bar{w}_k}.$$

Hence it is easy to see that

$$\frac{\partial \bar{w}_j}{\partial z_k}(0, \zeta) = \bar{\zeta} \sum_{\ell=1}^{n-1} \frac{\partial^2 g}{\partial w_j \partial w_\ell} \frac{\partial w_\ell}{\partial z_k}(0, \zeta)$$

and

$$\frac{\partial w_j}{\partial z_k}(0, \zeta) - \delta_{jk} = \zeta \sum_{\ell=1}^{n-1} \frac{\partial^2 \bar{g}}{\partial \bar{w}_j \partial \bar{w}_\ell} \frac{\partial \bar{w}_\ell}{\partial z_k}(0, \zeta).$$

Hence

$$\frac{\partial w_j}{\partial z_k}(0, \zeta) - \delta_{jk} = |\zeta|^2 \sum_{\ell,m} G_{j\ell} \overline{G_{\ell m}} \frac{\partial w_m}{\partial z_k}(0, \zeta).$$

Therefore

$$\left(\frac{\partial^2 \delta^2}{\partial z_j \partial \bar{z}_k}(0, \zeta) \right)_{1 \leq j,k \leq n-1} = I - (I - \bar{G}G)^{-1} = -\bar{G}G (I - \bar{G}G)^{-1}. \qquad (5.10)$$

Since

$$\delta(0,0) = |\zeta|, \quad \frac{\partial \delta^2}{\partial z_j}(0,\zeta) = 0 \quad \text{and} \quad \frac{\partial^2(-\log\delta)}{\partial z_j \partial \bar{z}_k} = \frac{1}{2}\left(-\frac{1}{\delta^2}\frac{\partial^2 \delta^2}{\partial w_j \partial \bar{w}_k} + \frac{1}{\delta^4}\frac{\partial \delta^2}{\partial z_j}\frac{\partial \delta^2}{\partial \bar{z}_k}\right),$$

the desired formula follows from (5.10). □

5.3.2 A Reduction Theorem in Tori

Let T be a complex torus of dimension n and let X be a connected Levi flat hypersurface of class C^ω in T. We shall say X is **holomorphically flat** if $T_X^{1,0} \subset$ Kerσ for some nonzero holomorphic 1-form σ on T. Since T is the quotient of $\mathbb{C}^n$ by the action of a lattice $\Gamma \subset \mathbb{C}^n$, the leaves of L_X for any holomorphically flat X are images of complex affine hyperplanes of $\mathbb{C}^n$ by the projection $\mathbb{C}^n \to T$. X is said to be **flat** if it is the image of an affine real hyperplane. We shall call a Kähler metric ω on T a **flat metric** if there exist holomorphic 1-forms $\sigma_1, \ldots, \sigma_n$ on T such that $\omega = i\sum_{j=1}^n \sigma_j \wedge \bar{\sigma}_j$. A **flat coordinate** around $x \in T$ is by definition a local coordinate $z = (z_1, \ldots, z_n)$ around x such that dz_j are extendible holomorphically to T as holomorphic 1-forms. Let us denote by $\delta_\omega(z)$ the distance from $z \in T$ to X with respect to ω. Since $\delta_\omega(z)$ is the infimum of the distances from z to the leaves of L_X, Lemma 5.3 implies the following:

Proposition 5.9. *X is holomorphically flat if and only if $\partial\bar{\partial} \log \delta_\omega$ has rank one near X.*

If X is holomorphically flat, it is clear that either X is flat or there exists a surjective holomorphic map from T to a complex torus of dimension one, say p, such that X is the preimage of some simple closed curve by p. Therefore, the remaining interest is to classify the rest. What is known at present is the following.

Theorem 5.13. *If X is not holomorphically flat, there exist a complex torus T' of dimension two, a surjective holomorphic map $\pi : T \to T'$, and a Levi flat hypersurface $X' \subset T'$ such that $\pi^{-1}(X') = X$.*

Proof. Let x be any point of X and let $\mathcal{L}$ be the leaf of L_X containing x. Let z be a flat coordinate of T around x such that $\mathcal{L} = \{z \; ; \; z_n = f(z')\}$ on a neighborhood of x for some holomorphic function f in $z' = (z_1, \ldots, z_{n-1})$ satisfying $df = 0$ at $z' = 0$. Then, with respect to the flat metric $\omega_x = i\sum_{j=1}^n dz_j \wedge d\bar{z}_j$,

$$\text{rank}\partial\bar{\partial} \log \delta_{\omega_x} \leq 2$$

near x. In fact, if rank$\partial\bar{\partial} \log \delta_{\omega_x}$ were strictly greater than two at points arbitrarily close to x, then by Lemma 5.3 one would have

$$\text{rank}\left(\frac{\partial^2 f}{\partial z_j \partial z_k}\right)_{1 \leq j,k \leq n-1} \geq 2$$

at $z' = 0$, which means by the real analyticity of X that the same would be true at almost all points of X with respect to some flat coordinates. This implies, by the real analyticity and linear algebra, that there exist flat metrics $\omega_1, \ldots, \omega_{2n-1}$ on T and a neighborhood $U \supset X$ such that $\sum_{j=1}^{2n-1} (-\log \delta_{\omega_j})$ is a plurisubharmonic function whose Levi form has everywhere at least three positive eigenvalues on $U \setminus X$. But this is impossible by Theorem 5.7. Similarly, Lemma 5.3 implies that $\partial\bar{\partial}(\log \delta_{\omega_1} + \log \delta_{\omega_2})$ has at most two nonzero eigenvalues near X, for any choice of flat metrics ω_1 and ω_2, and that the zero-eigenspace of $\partial\bar{\partial}\log \delta_{\omega_x}$ is equal to (the parallel translates of) Ker F at $(0, \zeta)$ for $0 < |\zeta| \ll 1$. Therefore Ker F must be actually all parallel as long as they are of dimension $n-2$. Since they are contained in the tangent spaces of the foliation extending L_X and X is not holomorphically flat, they have to be tangent to complex subtori of codimension two. They are the fibers of the desired fibration over T'. $\qquad\square$

Corollary 5.2. *Real analytic Levi flat hypersurfaces in a complex torus without nonconstant meromorphic functions are flat.*

Proof. Clearly, it suffices to prove that they are holomorphically flat. If they are not holomorphically flat, then take a reduction X' as above, which is also real analytic. If X' were not holomorphically flat, the Gauss map from X' associating to x the tangent space of $L_{X'}$ at x is a nonconstant map to the Riemann sphere, which extends holomorphically to a neighborhood of X' and hence meromorphically to T' because $T' \setminus X'$ turns out to be Stein by Lemma 5.3. (As for the Hartogs–type extension theorem for meromorphic functions, see [Siu-2, Chapter 1] and [M-P] for instance.) $\qquad\square$

The argument in the proof of Theorem 5.13 was originally used to prove the following more general assertion.

Theorem 5.14 (cf. [Oh-24]). *Let T be a complex torus, let $A \subset T$ be a closed set, and let F be a singular holomorphic foliation of codimension one on a neighborhood U of A such that A is a stable set of F. Suppose that the defining sheaf $\mathcal{F}$ of F is locally generated by a closed 1-form and topologically trivial on U. Then, either $\mathcal{F}$ is generated by a holomorphic 1-form on T or there exist a two-dimensional complex torus T' and a surjective holomorphic map $\pi : T \to T'$ such that $A = \pi^{-1}(\pi(A))$. Moreover, if A is not a complex analytic subset of T, $F = \pi^{-1}(F')$ holds on a neighborhood of A for some singular holomorphic foliation F' on a neighborhood of $\pi(A)$.*

Concerning singular holomorphic foliations of codimension one on complex tori, M. Brunella [Br-3] proved the following.

Theorem 5.15. *Let F be a singular holomorphic foliation of codimension one on a complex torus T such that $SingF \neq \emptyset$. Then there exist a complex torus T', a surjective holomorphic map $\pi : T \to T'$, and a singular holomorphic foliation F' on T' such that $\pi^{-1}(F') = F$ and $N_{F'}$ is positive.*

Remark 5.5. There exist plenty of singular holomorphic foliations of codimension one on any algebraic complex torus whose normal bundle is positive, for instance those induced from that on $\mathbb{CP}^n$ by a branched covering map. It is easy to see that the normal bundle of a (nonsingular) holomorphic foliation of codimension one on a complex torus is never positive.

Although our knowledge on Levi flat hypersurfaces in complex tori is generally incomplete because of the lack of classification in dimension two, we have still a complete classification at least when the torus has no nonconstant meromorphic functions (cf. Corollary 5.2). Therefore it is natural to expect that Levi flat hypersurfaces in non-algebraic surfaces can be classified similarly. In the case of Hopf surfaces, all the real analytic Levi flat hypersurfaces will be described below.

5.3.3   Classification in Hopf Surfaces

Hopf manifolds introduced by H. Hopf [Hf] are most typical non-Kähler manifolds. By definition, they are compact complex manifolds of dimension $n \geq 2$ whose universal covering space is $\mathbb{C}^n \setminus \{0\}$. We shall classify C^ω Levi flat hypersurfaces in Hopf surfaces based on the classification of Hopf surfaces ($n = 2$) by Kodaira [K-4].

Let $\mathscr{H}$ be a Hopf surface and let $\pi : \mathbb{C}^2 \setminus \{0\} \to \mathscr{H}$ be the universal covering. By analyzing the group $Gal(\mathscr{H}, \pi) := \{\sigma \in Aut(\mathbb{C}^2 \setminus \{0\}); \pi \circ \sigma = \pi\}$, Kodaira proved that every Hopf surface is isomorphic to the quotient of $\mathscr{H}$ with $Gal(\mathscr{H}, \pi) \cong \mathbb{Z}$ by a fixed point free action of a finite group. $\mathscr{H}$ is called **primary** if $Gal(\mathscr{H}, \pi) \cong \mathbb{Z}$, and **of diagonal type** if one can find $\tau \in Aut(\mathbb{C}^2 \setminus \{0\})$ such that $Gal(\mathscr{H}, \pi \circ \tau)$ is generated by the transformation $(z, w) \to (\alpha z, \beta w)$ for some $\alpha, \beta \in \mathbb{D} \setminus \{0\}$. It is known that a Hopf surface of diagonal type admits a nonconstant meromorphic function if and only if $\alpha^j = \beta^k$ holds for some $j, k \in \mathbb{N}$. In general, for a primary Hopf surface $\mathscr{H}$, Kodaira proved, applying a normalization due to Lattes [L], that one can find τ such that a generator of $Gal(\mathscr{H}, \pi \circ \tau)$ is given by

$$(z, w) \to \chi(z, w) = (\alpha z + \lambda w^m, \beta w),$$

where $0 < |\alpha| \leq |\beta| < 1$ and either $\lambda = 0$ or $\alpha = \beta^m$. Although the choice of $(\alpha, \beta, \lambda, m)$ is not unique for one $\mathscr{H}$, we shall denote the primary Hopf surfaces by $\mathscr{H}(\alpha, \beta, \lambda, m)$, and the associated covering map $\mathbb{C}^2 \setminus \{0\} \to \mathscr{H}(\alpha, \beta, \lambda, m)$ by π. Kodaira [K-4] proved that every primary Hopf surface is diffeomorphic to $S^1 \times S^3$. Diffeomorphism types of general Hopf surfaces were classified by M. Kato [Ka]. As a basic complex analytic property of Hopf surfaces one has the following.

Proposition 5.10. *A Hopf surface with a nonconstant meromorphic function is holomorphically mapped onto $\mathbb{CP}^1 (= \hat{\mathbb{C}})$.*

Proof. Let f be a nonconstant meromorphic function on $\mathscr{H}$ and let $C = f^{-1}(0)$. Then C does not intersect with $f^{-1}(1)$. In fact, had it not been the case, then the self–intersection number of C would be positive, so that the transcendence degree of the field of meromorphic functions over $\mathbb{C}$ would be two (cf. Theorem 2.45 for instance), which is absurd because it would imply that $\mathscr{H}$ is projective algebraic by the classical Chow–Kodaira theorem (cf. [C-K]). Therefore $\mathscr{H}$ is holomorphically mapped onto $\mathbb{CP}^1$ by f. ☐

Remark 5.6. It is easy to see that a primary Hopf surface with a nonconstant meromorphic function is of diagonal type.

Let X be a real analytic Levi flat hypersurface in a primary Hopf surface $\mathscr{H}$. We shall give explicit descriptions for X by assuming that $\mathscr{H}$ is primary, putting aside a question which X is invariant under finite group actions. Let us first assume that $\mathscr{H} \setminus X$ is Stein. In this situation, the section of the projectivization $\mathbb{P}(T^{1,0}_{\mathscr{H}})$ over X induced by $T^{1,0}_{L_X}$ is extended to $\mathscr{H}$ as a meromorphic section, say h. It was observed in [Oh-30] that X can be described by studying the intersection of $h(\mathscr{H})$ and the levels of a nonconstant meromorphic function on $\mathbb{P}(T^{1,0}\mathscr{H})$. For that, the first step is of course the following.

Proposition 5.11. $\mathbb{P}(T^{1,0}_{\mathscr{H}})$ *admits a nonconstant meromorphic function.*

Proof. Let $\chi(z, w)$ be as above. Then $T^{1,0}_{\mathscr{H}}$ is equivalent to the quotient of $(\mathbb{C}^2 \setminus \{0\}) \times \mathbb{C}^2$ by the action of the group generated by $(\chi, d\chi)$. Letting $((z, w), (\xi, \eta))$ be the coordinate of $\mathbb{C}^2 \times \mathbb{C}^2$ so that (ξ, η) represents the vector $\xi \frac{\partial}{\partial z} + \eta \frac{\partial}{\partial w}$, one has

$$d\chi(\xi, \eta) = (\alpha\xi + m\lambda w^{m-1}\eta)\frac{\partial}{\partial z} + \beta\eta\frac{\partial}{\partial w}.$$

Therefore $\frac{\xi w}{\eta z}$ is invariant under $(\chi, d\chi)$ if $\lambda = 0$, and so is the function

$$\frac{\xi}{\eta w^{m-1}} - \frac{mz}{w^m}$$

otherwise.   ☐

Applying this, C^ω Levi flat hypersurfaces in primary Hopf surfaces can be specified. In order to give the description of these, let us recall the construction of Nemirovski in [Nm] in this situation.

Let $p : \mathbb{C}^2 \setminus \{0\} \to \mathbb{CP}^1$ be the natural projection. Let $\zeta = z/w$ be the inhomogeneous coordinate of $\mathbb{CP}^1$ and let

$$U_+ = \{\zeta; 0 \le |\zeta| < \infty\},$$
$$U_- = \{\zeta; 0 < |\zeta| \le \infty\}.$$

Let ω_+ and ω_- respectively be meromorphic 1-forms on U_+ and U_- satisfying

$$\omega_+ - \omega_- = d\log\zeta \quad on \quad U_+ \cap U_-.$$

Then, parallel transports of the points in $\mathbb{C}^2 \setminus \{0\}$ are defined over the paths avoiding the poles of $\omega_{\pm}$. In terms of the fiber coordinates w, z of $p^{-1}(U_+)$, $p^{-1}(U_-)$, the parallel transport along $\gamma : [0, 1] \to U_{\pm}$ is given by

$$w \to w e^{\int_\gamma \omega_+} \quad \text{on} \quad p^{-1}(U_+)$$

and

$$z \to z e^{\int_\gamma \omega_-} \quad \text{on} \quad p^{-1}(U_-).$$

Let P_∞ be the union of the sets of poles of ω_+ and ω_-. Then, for any point $\zeta_0 \in \mathbb{CP}^1 \setminus P_\infty$, a closed real analytic curve C in $p^{-1}(\zeta_0)$ yields a Levi flat hypersurface in $(\mathbb{C}^2 \setminus \{0\}) \setminus p^{-1}(P_\infty)$ as long as the parallel transport along a curve γ as above with $\gamma(0) = \gamma(1) = \zeta_0$ with respect to $\omega_{\pm}$ leave C invariant. In such a case, if moreover the closure of the union of the parallel transports of C in $\mathbb{C}^2 \setminus \{0\}$ is a smooth hypersurface, we shall call it a Levi flat hypersurface of Nemirovski type since it is an analogue of $\bar{S} \subset E^*$ in Sect. 5.1.2.

Theorem 5.16. *Let X be a real analytic Levi flat hypersurface with Stein complement in a Hopf surface $\mathcal{H}(\alpha, \beta, \lambda, m)$. If $\lambda = 0$ or $m = 1$, then the preimage of X by the covering map π is of Nemirovski type.*

Proof. Let h be a meromorphic section of the bundle $q : \mathbb{P}(T_{\mathcal{H}}^{1,0}) \to \mathcal{H}$ induced by $T_{Lx}^{1,0}$. If $\lambda = 0$, then ξ/η is constant on $h(\mathcal{H})$, because otherwise there would be nonconstant meromorphic functions on $\mathcal{H}$ and $h(\mathcal{H}) \cap (\xi w/\eta z)^{-1}(c)$ are mapped by q to complex curves in $\mathcal{H}$ which intersect with the images of $z = 0$ and $w = 0$. This contradicts the Cho-Kodaira theorem. Therefore, there exists a meromorphic function c in z/w such that the vector field $c\frac{\partial}{\partial z} + \frac{\partial}{\partial w}$ or $\frac{\partial}{\partial z} + \frac{1}{c}\frac{\partial}{\partial w}$ is everywhere tangent to $\pi^{-1}(X)$. Hence $\pi^{-1}(X)$ is of Nemirovski type. If $m = 1$, a similar argument applies to the function $(\xi/\eta - z/w)/(z/w - \lambda)$ instead of $\xi w/\eta z$. $\square$

A similar method works for the case $\lambda \neq 0$ and $m \geq 2$ to describe a C^ω Levi flat hypersurface X with Stein complement in terms of the holomorphic map

$$\tilde{p} : \mathbb{C}^2 \setminus \{0\} \to \mathbb{CP}^1,$$

where

$$\tilde{p}(z, w) = \left(z + \frac{aw^m}{m} : w\right)$$

and a is a constant such that the image of h is contained in the preimage of a by $\xi/\eta w^{m-1} - mz/w^m$. Namely, the preimage of X by $\tilde{p}$ is "of generalized Nemirovski type" (cf. [Oh-30]).

The following is due to Kim, Levenberg and Yamguchi [K-L-Y] and Levenberg and Yamaguchi [L-Y-2]. (See also Miebach [Mb].)

Theorem 5.17. *Let X be a real analytic Levi flat hypersurface in a Hopf surface of diagonal type* $\mathcal{H} = \mathcal{H}(\alpha, \beta, 0, 0)$. *If* $\mathcal{H} \setminus X$ *is not Stein, then X is either of the form*

$$k|z|^{\frac{\log |\alpha|}{\log |\beta|}} = |w| \quad (k > 0)$$

or the preimage of a Jordan curve in $\mathbb{CP}^1$ *by a surjective holomorphic map.*

Recently, Theorem 5.17 was complemented by the following.

Theorem 5.18 (cf. [Oh-31]). *A primary Hopf surface is of diagonal type if and only if it contains a real analytic Levi flat hypersurface whose complement is not Stein.*

For the proof, Lemma 5.2 is crucial.

Bibliography

[A-B]   Adachi, M., Brinkschulte, J.: Curvature restrictions for Levi-flat real hypersurfaces in complex projective planes, to appear in Ann. de l'Inst. Fourier

[A]   Ahlfors, L.V.: Conformal Invariants, Topics in Geometric Function Theory, xii+162 pp. Reprint of the 1973 original. With a foreword by Peter Duren, F. W. Gehring and Brad Osgood. AMS Chelsea Publishing, Providence (2010)

[A-Bl] Ahlfors, L.V., Beurling, A.: Conformal invariants and function-theoretic null-sets. Acta Math. **33**, 105–129 (1950)

[A-N]   Akizuki, Y., Nakano, S.: Note on Kodaira-Spencer's proof of Lefschetz theorems. Proc. Jpn. Acad. **30**, 266–272 (1954)

[A-Gh] Andreotti, A., Gherardelli, F.: Some remarks on quasi-abelian manifolds. In: Global Analysis and Its Applications, vol. II, pp. 203–206. International Atomic Energy Agency, Vienna (1974)

[A-G]   Andreotti, A., Grauert, H.: Théorème de finitude pour la cohomologie des espaces complexes. Bull. Soc. Math. Fr. **90**, 193–259 (1962)

[A-V-1] Andreotti, A., Vesentini, E.: Sopra un teorema di Kodaira. Ann. Scuola Norm. Sup. Pisa **15**, 283–309 (1961)

[A-V-2] Andreotti, A., Vesentini, E.: Carleman estimates for the Laplace-Beltrami equation on complex manifolds. Inst. Hautes Études Sci. Publ. Math. **25**, 81–130 (1965)

[A-S]   Angehrn, U., Siu, Y.-T.: Effective freeness and point separation for adjoint bundles. Invent. Math. **122**, 291–308 (1995)

[Ar]   Aronszajn, N.: A unique continuation theorem for solutions of elliptic partial differential equations or inequalities of second order. J. Math. Pures Appl. **36**, 235–249 (1957)

[B-1]   Barlet, D.: Espace analytique réduit des cycles analytiques complexes compacts d'un espace analytique complexe de dimension finie. In: Fonctions de plusieurs variables complexes, II (Sém. Franois Norguet, 1974–1975). Lecture Notes in Mathematics, vol. 482, pp. 1–158. Springer, Berlin (1975)

[B-2]   Barlet, D.: Convexité de l'espace des cycles. Bull. Soc. Math. Fr. **106**, 373–397 (1978)

[B-B-B] Battaglia, E., Biagi, S., Bonfiglioli, A.: The strong maximum principle and the Harnack inequality for a class of hypoelliptic non-Hörmander operators, preprint

[By]   Bayer, D.: The division algorithm and the Hilbert scheme. Ph.D. thesis, Harvard University (1982)

[B-T-1] Bedford, E., Taylor, B.A.: Variational properties of the complex Monge-Ampère equation I. Dirichlet principle. Duke Math. J. **45**, 375–403 (1978)

© Springer Japan 2015   177
T. Ohsawa, L^2 *Approaches in Several Complex Variables*, Springer Monographs in Mathematics, DOI 10.1007/978-4-431-55747-0

[B-T-2]   Bedford, E., Taylor, B.A.: Variational properties of the complex Monge-Ampère equation II. Intrinsic norms. Am. J. Math. **101**, 1131–1166 (1979)

[B-S]   Behnke, H., Stein, K.: Entwicklung analytischer Funktionen auf Riemannschen Flächen. Math. Ann. **120**, 430–461 (1949)

[B-T]   Behnke, H., Thullen, P.: Theorie der Funktionen mehrerer komplexer Veränderlichen, Ergebnisse der Mathematik und ihrer Grenzgebiete. (2. Folge) (by Heinrich Behnke, P. Thullen, R. Remmert, W. Barth, O. Forster, W. Kaup, H. Kerner, H. Holmann, H.J. Reiffen, G. Scheja, K. Spallek) Ergebnisse der Mathematik und ihrer Grenzgebiete. 2. Folge (Book 51) Springer, 1970. (Softcover reprint : October 20, 2011)

[B]   Behrens, M.: Plurisubharmonic defining functions of weakly pseudoconvex domains in $\mathbb{C}^2$. Math. Ann. **270**, 285–296 (1985)

[B-G-V-Y] Berenstein, C.A., Gay, R., Vidras, A., Yger, A.: Residue Currents and Bezout Identities. Progress in Mathematics, vol. 114, xii+158 pp. Birkhäuser Verlag, Basel (1993)

[Be]   Bergman, S.: Über die entwicklung der harmonischen Funktionen der Ebene und des Raumes nach Orthogonalfunktionen. Math. Ann. **86**, 238–271 (1922)

[B-B-S]   Berman, R., Berndtsson, B., Sjöstrand, J.: A direct approach to Bergman kernel asymptotics for positive line bundles. Ark. Mat. **46**, 197–217 (2008)

[Brd-1]   Berndtsson, B.: Curvature of vector bundles associated to holomorphic fibrations. Ann. Math. **169**, 531–560 (2009)

[Brd-2]   Berndtsson, B.: The openness conjecture for plurisubharmonic functions. arXiv: 1305.5781

[Brd-L]   Berndtsson, B., Lempert, L.: A proof of the Ohsawa-Takegoshi theorem with sharp estimates, arXiv: 140/.4946v1 [math.CV], 18 July 2014, to appear in J. Math. Soc. Jpn.

[B-Pa]   Berndtsson, B., Păun, M.: Bergman kernels and the pseudoeffectivity of relative canonical bundles. Duke Math. J. **145**, 341–378 (2008)

[B-M-1]   Bierstone, E., Milman, P.; Relations among analytic functions I. Ann. Inst. Fourier **37**, 187–239 (1987)

[B-M-2]   Bierstone, E., Milman, P.: Uniformization of analytic spaces. J. Am. Math. Soc. **2**, 801–836 (1989)

[B-L]   Birkenhake, C., Lange, H.: Complex Abelian Varieties. Grundlehren der Mathematischen Wissenschaften, vol. 302, 2nd edn., xii+635 pp. Springer, Berlin (2004)

[Bi]   Bishop, E.: Mappings of partially analytic spaces. Am. J. Math. **83**, 209–242 (1961)

[Bł-1]   Błocki, Z.: The Bergman metric and the pluricomplex Green function. Trans. Am. Math. Soc. **357**, 2613–2625 (2005)

[Bł-2]   Błocki, Z.: Suita conjecture and the Ohsawa-Takegoshi extension theorem. Invent. Math. **193**, 149–158 (2013)

[Bł-3]   Błocki, Z.: A lower bound for the Bergman kernel and the Bourgain-Milman inequality. In: Klartag, B., Milman, E. (eds.) Geometric Aspects of Functional Analysis, Israel Seminar (GAFA) 2011–2013. Lecture Notes in Mathematics, vol. 2116, pp. 53–63. Springer (2014)

[B-P]   Błocki, Z., Pflug, P.: Hyperconvexity and Bergman completeness. Nagoya Math. J. **151**, 221–225 (1998)

[Bo]   Bochner, S.: Über orthogonale Systeme analytischer Funktionen. Math. Z. **14**, 180–207 (1922)

[Bb-1]   Bombieri, E.: Algebraic values of meromorphic maps. Invent. Math. **10**, 267–287 (1970). Addendum: Invent. Math. **11**, 163–166 (1970)

[Bb-2]   Bombieri, E.: The pluricanonical map of a complex surface. In: Several Complex Variables, I. Proceedings of the Conference on University of Maryland, College Park, 1970, pp. 35–87. Springer, Berlin (1970)

[B-F-J]   Boucksom, S., Favre, C., Jonsson, M.: Valuations and plurisubharmonic singularities. Publ. RIMS **44**, 449–494 (2008)

[Bou]   Bouligand, G.: Sur les problème de Dirichlet. Ann. Soc. Polonaise de Math.
          4, 59–112 (1925)
[Brm]   Bremermann, H.J.: On the conjecture of the equivalence of the plurisubharmonic
          functions and the Hartogs functions. Math. Ann. **131**, 76–86 (1956)
[B-Sk]   Briancon, J., Skoda, H.: Sur la clôture intégrale d'un idéal de germes de fonctions
          holomorphes en un point de $\mathbb{C}^n$. C. R. Acad. Sci. Paris Sér. A **278**, 949–951 (1974)
[Br-1]   Brunella, M.: A positivity property for foliations on compact Kähler manifolds. Int.
          J. Math. **17**, 35–43 (2006)
[Br-2]   Brunella, M.: On the dynamics of codimension one holomorphic foliations with
          ample normal bundle. Indiana Univ. Math. J. **57**, 3101–3113 (2008)
[Br-3]   Brunella, M.: Codimension one foliations on complex tori. Ann. Fac. Sci. Toulouse
          Math. **19**, 405–418 (2010)
[Br-4]   Brunella, M.: Birational Geometry of Foliations. IMPA Monographs, vol. 1, 124 p.
          Springer, Cham (2015)
[C-LN-S] Camacho, C., Lins Neto, A., Sad, P.: Minimal sets of foliations on complex projective
          spaces. Inst. Hautes Études Sci. Publ. Math. **68**, 187–203 (1988)
[CJ-1]   Cao, J.-Y.: Numerical dimension and a Kawamata-Viehweg-Nadel-type vanishing
          theorems on compact Kähler manifolds. Compos. Math. **150**, 1869–1902 (2014)
[CJ-2]   Cao, J.-Y.: Ohsawa-Takegoshi extension theorem for compact Kähler manifolds and
          applications. arXiv: 1404.6937
[C-S]   Cao, J., Shaw, M.-C.: The $\bar{\partial}$-Cauchy problem and nonexistence of Lipschitz Levi-flat
          hypersurfaces in $\mathbb{CP}^n$ with $n \geq 3$. Math. Z. **256**, 175–192 (2007)
[C-C]   Capocasa, F., Catanese, F.: Periodic meromorphic functions. Acta Math. **166**, 27–68
          (1991)
[Ca]   Carleson, L.: Selected Problems on Exceptional Sets. Van Nostrand Mathematical
          Studies, No. 13, v+151 pp. D. Van Nostrand, Princeton/Toronto/London (1967)
[C]   Cartan, H.: Idéaux et modules de fonctions analytiques de variables complexes. Bull.
          Soc. Math. Fr. **78**, 29–64 (1950)
[C-T]   Cartan, H., Thullen, P.: Zur Theorie der Singularitäten der Funktionen mehrerer
          komplexen Veränderlichen: Regularitäts– und Konvergenz–bereiche. Math Ann.
          106, 617–647 (1932)
[Ct]   Catlin, D.: The Bergman kernel and a theorem of Tian. In: Analysis and Geometry
          in Several Complex Variables, Katata, 1997. Trends in Mathematics, pp. 1–23.
          Birkhäuser, Boston (1999)
[Cv]   Cerveau, D.: Minimaux des feuilletages algébriques de $\mathbb{CP}^n$. Ann. Inst. Fourier
          (Grenoble) **43**, 1535–1543 (1993)
[C-LN]   Cerveau, D., Lins Neto, A.: A structural theorem for codimension-one foliations on
          $\mathbb{P}^n$, $n \geq 3$, with an application to degree-three foliations. Ann. Sc. Norm. Super. Pisa
          12, 1–41 (2013)
[C-G-M]   Cheeger, J., Goresky, M., MacPherson, R.: L^2-cohomology and intersection homol-
          ogy of singular algebraic varieties. In: Yau, S.-T. (ed.) Seminar on Differential
          Geometry, pp. 303–340. Princeton University Press, Princeton (1982)
[Ch-1]   Chen, B.-Y.: Bergman completeness of hyperconvex manifolds. Nagoya Math. J.
          175, 165–170 (2004)
[Ch-2]   Chen, B.-Y.: A simple proof of the Ohsawa-Takegoshi extension theorem.
          arXiv:math.CV/. 1105.2430
[C-K]   Chow, W.-L., Kodaira, K.: On analytic surfaces with two independent meromorphic
          functions. Proc. Natl. Acad. Sci. U.S.A. **38**, 319–325 (1952)
[Ch]   Chow, W.-L.: On compact complex analytic varieties. Am. J. Math. **71**, 893–914
          (1949)
[C-L]   Coeuré, G., Loeb, J.J.: A counterexample to the Serre problem with a bounded
          domain of $\mathbb{C}^2$ as fiber. Ann. Math. **122**, 329–334 (1985)
[dC]   de Cataldo, M.A.A.: Singular Hermitian metrics on vector bundles. J. Reine Angew.
          Math. **502**, 93–122 (1998)

[Dm-1]   Demailly, J.-P.: Un example de fibré holomorphe non de Stein à fibre $\mathbb{C}^2$ ayant pour base le disque ou le plan. Invent. Math. **48**, 293–302 (1978)

[Dm-2]   Demailly, J.-P.: Estimations L^2 pour l'opérateur $\bar{\partial}$ d'un fibré vectoriel holomorphe semi-positif au-dessus d'une variété kählérienne complète. Ann. Sci. École Norm. Sup. **15**, 457–511 (1982)

[Dm-3]   Demailly, J.-P.: Sur l'identité de Bochner-Kodaira-Nakano en géométrie hermitienne. Lecture Notes in Mathematics, vol. 1198, pp. 88–97. Springer, Berlin/Heidelberg (1985)

[Dm-4]   Demailly, J.-P.: Une généralisation du théorème d'annulation de Kawamata-Viehweg. C. R. Acad. Sci. Paris Sér. I Math. **309**, 123–126 (1989)

[Dm-5]   Demailly, J.-P.: Cohomology of q-convex spaces in top degrees. Math. Z. **204**, 283–295 (1990)

[Dm-6]   Demailly, J.-P.: Regularization of closed positive currents and intersection theory. J. Algebr. Geom. **1**, 361–409 (1992)

[Dm-7]   Demailly, J.-P.: On the Ohsawa-Takegoshi-Manivel L^2 extension theorem. In: Complex Analysis and Geometry. Progress in Mathematics, vol. 188, pp. 47–82. Birkhäuser, Basel/Boston (2000)

[Dm-8]   Demailly, J.-P.: Multiplier ideal sheaves and analytic methods in algebraic geometry. In: School on Vanishing Theorems and Effective Results in Algebraic Geometry (Trieste, 2000). ICTP Lecture Notes, vol. 6, pp. 1–148. Abdus Salam International Centre for Theoretical Physics, Trieste (2001)

[Dm-9]   Demailly, J.-P.: Analytic Methods in Algebraic Geometry. Surveys of Modern Mathematics, vol. 1, viii+231 pp. International Press/Higher Education Press, Somerville/Beijing (2012)

[Dm-E-L]   Demailly, J.-P., Ein, L., Lazarsfeld, R: A subadditivity property of multiplier ideals. Dedicated to William Fulton on the occasion of his 60th birthday. Mich. Math. J. **48**, 137–156 (2000)

[Dm-K]   Demailly, J.-P., Kollár, J.: Semi-continuity of complex singularity exponents and Kähler-Einstein metrics on Fano orbifolds. Ann. Sci. École Norm. Sup. **34**, 525–556 (2001)

[Dm-P]   Demailly, J.-P., Peternell, T.: A Kawamata-Viehweg vanishing theorem on compact Kähler manifolds. J. Differ. Geom. **63**, 231–277 (2003)

[Dm-P-S]   Demailly, J.-P., Peternell, T., Schneider, M.: Pseudo-effective line bundles on compact Kähler manifolds. Int. J. Math. **12**, 689–741 (2001)

[Dm-S]   Demailly, J.-P., Skoda, H.: Relations entre les notions de positivité de P.A. Griffiths et de S. Nakano, Seminaire P. Lelong-H. Skoda (Analyse), année 1978/79. Lecture Notes in Mathematics, vol. 822, pp. 304–309. Springer, Berlin (1980)

[D]   Diederich, K.: Das Randverhalten der Bergmanschen Kernfunktion und Metrik in streng pseudo-konvexen Gebieten. Math. Ann. **187**, 9–36 (1970)

[D-F-1]   Diederich, K., Fornaess, J.E.: Pseudoconvex domains: bounded strictly plurisubharmonic exhaustion functions. Invent. Math. **39**, 129–141 (1977)

[D-F-2]   Diederich, K., Fornaess, J.E.: Smooth, but not complex-analytic pluripolar sets. Manuscr. Math. **37**, 121–125 (1982)

[D-F-3]   Diederich, K., Fornaess, J.E.: A smooth curve in $\mathbb{C}^2$ which is not a pluripolar set. Duke Math. J. **49**, 931–936 (1982)

[D-F-4]   Diederich, K., Fornaess, J.E.: Thin complements of complete Kähler domains. Math. Ann. **259**, 331–341 (1982)

[D-F-5]   Diederich, K., Fornaess, J.E.: A smooth pseudoconvex domain without pseudoconvex exhaustion. Manuscr. Math. **39**, 119–123 (1982)

[D-F-6]   Diederich, K., Fornaess, J.E.: On the nature of thin complements of complete Kähler metrics. Math. Ann. **268**, 475–495 (1984)

[D-Oh-1]   Diederich, K., Ohsawa, T.: A Levi problem on two-dimensional complex manifolds. Math. Ann. **261**, 255–261 (1982)

[D-Oh-2]   Diederich, K., Ohsawa, T.: Harmonic mappings and disc bundles over compact Kähler manifolds. Publ. Res. Inst. Math. Sci. **21**, 819–833 (1985)

[D-Oh-3]   Diederich, K., Ohsawa, T.: On the parameter dependence of solutions to the $\bar{\partial}$-equation. Math. Ann. **289**, 581–587 (1991)

[D-Oh-4]   Diederich, K., Ohsawa, T.: An estimate for the Bergman distance on pseudoconvex domains. Ann. Math. **141**, 181–190 (1995)

[D-P]   Diederich, K., Pflug, P.: Über Gebiete mit vollständiger Kählermetrik. Math. Ann. **257**, 191–198 (1981)

[Ds]   Donaldson, S.K.: Scalar curvature and projective embeddings I. J. Differ. Geom. **59**, 479–522 (2001)

[D-F]   Donnelly, H., Fefferman, C.: L^2-cohomology and index theorem for the Bergman metric. Ann. Math. **118**, 593–618 (1983)

[E-L]   Ein, L., Lazarsfeld, R.: Global generation of pluricanonical and adjoint linear series on smooth projective threefolds. J. Am. Math. Soc. **6**, 875–903 (1993)

[Eb]   Eisenbud, D.: Commutative Algebra with a View Toward Algebraic Geometry. Graduate Texts in Mathematics, vol. 150. Springer, New York (1995)

[Eb-E]   Eisenbud, D., Evans, E.G., Jr.: Every algebraic set in n-space is the intersection of n hypersurfaces. Invent. Math. **19**, 107–112 (1973)

[E]   Elencwajg, G.: Pseudo-convexité locale dans les variétés kählériennes. Ann. Inst. Fourier (Grenoble) **25**, 295–314 (1975)

[E-Grm]   Eliashberg, Y., Gromov, M.: Embeddings of Stein manifolds of dimension n into the affine space of dimension 3n/2+1. Ann. Math. **136**, 123–135 (1992)

[E-K-P-R]   Eyssidieux, P., Katzarkov, L., Pantev, T., Ramachandran, M.: Linear Shafarevich conjecture. Ann. Math. **176**, 1545–1581 (2012)

[F-J]   Favre, C., Jonsson, M.: Valuations and multiplier ideals. J. Am. Math. Soc. **18**, 655–684 (2005)

[F]   Fefferman, C.: Monge-Ampère equations, the Bergman kernel, and geometry of pseudoconvex domains. Ann. Math. **103**, 395–416 (1976)

[Fj-1]   Fujiki, A.: On the blowing down of analytic spaces. Publ. Res. Inst. Math. Sci. **10**, 473–507 (1974/1975)

[Fj-2]   Fujiki, A.: An L^2 Dolbeault lemma and its applications. Publ. Res. Inst. Math. Sci. **28**, 845–884 (1992)

[F-N]   Fujiki, A., Nakano, S.: Supplement to "On the inverse of monoidal transformation". Publ. Res. Inst. Math. Sci. **7**, 637–644 (1971/1972)

[FR]   Fujita, R.: Domaines sans point critique intérieur sur l'espace projectif complexe. J. Math. Soc. Jpn. **15**, 443–473 (1963)

[F-1]   Fujita, T.: On Kähler fiber spaces over curves. J. Math. Soc. Jpn. **30**, 779–794 (1978)

[F-2]   Fujita, T.: On polarized manifolds whose adjoint bundles are not semipositive. In: Algebraic Geometry, Sendai, 1985. Advanced Studies in Pure Mathematics, vol. 10, pp. 167–178 (1987)

[F-3]   Fujita, T.: Remarks on Ein-Lazarsfeld criterion of spannedness of adjoint bundles of polarized threefolds, preprint 1993

[Ga]   Gaffney, M.P.: A special Stokes's theorem for complete Riemannian manifolds. Ann. Math. **60**, 140–145 (1954)

[G]   Galligo, A.: Théorème de division et stabilité en géométrie analytique locale. Ann. Inst. Fourier **29**, 107–184 (1979)

[Gh]   Ghys, É.: Feuilletages holomorphes de codimension un sur les espaces homogènes complexes. Ann. Fac. Sci. Toulouse Math. **5**, 493–519 (1996)

[G-Km]   Gong, X.-H., Kim, K.-T.: The $\bar{\partial}$-equation on variable strictly pseudoconvex domains. Preprint

[Gra-1]   Grauert, H.: Charakterisierung der holomorph vollständigen komplexen Räume. Math. Ann. **129**, 233–259 (1955)

[Gra-2]   Grauert, H.: Charakterisierung der Holomorphiegebiete durch die vollständige Kählersche Metrik. Math. Ann. **131**, 38–75 (1956)

[Gra-3]   Grauert, H.: On Levi's problem and the imbedding of real-analytic manifolds. Ann.
            Math. **68**, 460–472 (1958)

[Gra-4]   Grauert, H.: Ein Theorem der analytischen Garbentheorie und die Modulräume
            komplexer Strulturen. Inst. H. E. S. Publ. Math. 233–292 (1960)

[Gra-5]   Grauert, H.: Über Modifikationen und exzeptionelle analytische Mengen. Math. Ann.
            146, 331–368 (1962)

[Gra-6]   Grauert, H.: Bemerkenswerte pseudokonvexe Mannigfaltigkeiten. Math. Z. **81**, 377–
            391 (1963)

[Gra-R-1]   Grauert, H., Remmert, R.: Theory of Stein Spaces (Translated from the German
            by Alan Huckleberry). Reprint of the 1979 translation. Classics in Mathematics,
            xxii+255 pp. Springer, Berlin (2004)

[Gra-R-2]   Grauert, H., Remmert, R.: Coherent Analytic Sheaves. Grundlehren der Mathematis-
            chen Wissenschaften, vol. 265, xviii+249 pp. Springer, Berlin (1984)

[Gra-Ri-1] Grauert, H., Riemenschneider, O.: Kählersche Mannigfaltigkeiten mit hyper-q-
            konvexem Rand. In: Problems in Analysis. Lectures Symposium in Honor of
            Salomon Bochner. Princeton University, Princeton, 1969, pp. 61–79. Princeton
            University Press, Princeton (1970)

[Gra-Ri-2] Grauert, H., Riemenschneider, O.: Verschwindungssätze für analytische Kohomolo-
            giegruppen auf komplexen Räumen. Invent. Math. **11**, 263–292 (1970)

[G-W-1]   Greene, R.E., Wu, H.-H.: Embedding of open Riemannian manifolds by harmonic
            functions. Ann. Inst. Fourier **25**, 215–235 (1975)

[G-W-2]   Greene, R.E., Wu, H.-H.: Integrals of subharmonic functions on manifolds of
            nonnegative curvature. Invent. Math. **27**, 265–298 (1974)

[G-W-3]   Greene, R.E., Wu, H.-H.: Some function-theoretic properties of noncompact Kähler
            manifolds. In: Differential Geometry. Proceedings of Symposia in Pure Mathematics,
            vol. XXVII, Part 2, Stanford University, Stanford, 1973, pp. 33–41. American
            Mathematical Society, Providence (1975)

[G-W-4]   Greene, R.E., Wu, H.-H.: Gap theorems for noncompact Riemannian manifolds.
            Duke Math. J. **49**, 731–756 (1982)

[Gri-1]   Griffiths, P.A.: Periods of integrals on algebraic manifolds. I and II. Am. J. Math. **90**,
            568–626 and 805–865 (1968)

[Gri-2]   Griffiths, P.A.: Hermitian differential geometry, Chern classes, and positive vector
            bundles. In: Global Analysis. Papers in Honor of K. Kodaira, pp. 185–251. University
            Tokyo Press, Tokyo (1969)

[Gri-3]   Griffiths, Ph.A.: Periods of integrals on algebraic manifolds. III, some global
            differential-geometric properties of the period mapping. Inst. Hautes Études Sci.
            Publ. Math. **38**, 125–180 (1970)

[Grm]   Gromov, M.: Kähler hyperbolicity and L^2-Hodge theory. J. Differ. Geom. **33**, 263–
            292 (1991)

[G-K]   Greene, R.E., Krantz, S.G.: The stability of the Bergman kernel and the geometry of
            the Bergman metric. Bull. Am. Math. Soc. **4**, 111–115 (1981)

[Grt-1]   Grothendieck, A.: La théorie des classes de Chern. Bull. Soc. Math. Fr. **86**, 137–154
            (1958)

[Grt-2]   Grothendieck, A.: Sur une note de Mattuck-Tate. J. Reine Angew. Math. **200**, 208–
            215 (1958)

[G-Z-1]   Guan, Q.-A., Zhou, X.-Y.: A solution of an L2 extension problem with optimal
            estimate and applications. Ann. Math. **181**, 1139–1208 (2015)

[G-Z-2]   Guan, Q.-A., Zhou, X.-Y.: Strong openness conjecture for plurisubharmonic func-
            tions (2013). arXiv:1311.3/81v1

[G-Z-3]   Guan, Q.-A., Zhou, X.-Y.: Strong openness conjecture and related problems for
            plurisubharmonic functions (2014). arXiv:1401.7158v1

[G-Z-4]   Guan, Q.-A., Zhou, X.-Y.: Classification of multiplier ideal sheaf with Lelong
            number one weight (2014). arXiv:1411.6737v1

[G-Z-5]   Guan, Q.-A., Zhou, X.-Y.: A proof of Demailly's strong openness conjecture, preprint

[G-R]   Gunning, R.C., Rossi, H.: Analytic Functions of Several Complex Variables, xiv+317 pp. Prentice-Hall, Englewood Cliffs (1965)

[Ha]   Hamano, S.: Variation formulas for L_1-principal functions and application to the simultaneous uniformization problem. Mich. Math. J. **60**, 271–288 (2011)

[H]   Hartogs, F.: Zur Theorie der analytischen Funktionen mehrerer unabhängiger Veränderlichen insbesondere über die Darstellung derselben durch Reihen, welche nach Potenzen einer Veränderlichen fortschreiten. Math. Ann. **62**, 1–88 (1906)

[Hei]   Heier, G.: Effective freeness of adjoint line bundles. Doc. Math. **7**, 31–42 (2002)

[H-1]   Helmke, S.: On Fujita's conjecture. Duke Math. J. **88**, 201–216 (1997)

[H-2]   Helmke, S.: On global generation of adjoint linear systems. Math. Ann. **313**, 635–652 (1999)

[He]   Henkin, G.M.: The method of integral representations in complex analysis. In: Several Complex Variables I. Encyclopaedia of Mathematical Sciences, vol. 7, pp. 19–116. Springer, Berlin/New York (1990). (The English translation of the article in) Current problems in mathematics. Fundamental directions, vol. 7, 23–124, **258**, Itogi Nauki i Tekhniki, Akad. Nauk SSSR, Vsesoyuz. Inst. Nauchn. i Tekhn. Inform., Moscow (1985)

[Hb]   Herbort, G.: The Bergman metric on hyperconvex domains. Math. Z. **232**, 183–196 (1999)

[Hp]   Hiep, P.H.: The weighted log canonical threshold (2014). arXiv:1401.4833v5

[Hn]   Hironaka, H.: Flattening theorem in complex-analytic geometry. Am. J. Math. **97**, 503–547 (1975)

[H-U]   Hironaka, H., Urabe, T.: Kaiseki kukan nyumon. (Japanese) (Introduction to analytic spaces), 2nd edn. Suri Kagaku Raiburari (Mathematical Science Library), 1, vi+158 pp. Asakura Publishing Co., Tokyo (1983)

[Hr]   Hirschowitz, A.: Pseudoconvexité au-dessus d'espaces plus ou moins homogènes. Invent. Math. **26**, 303–322 (1974)

[Ho]   Hodge, W.V.D.: The Theory and Applications of Harmonic Integrals. Cambridge University Press, Cambridge (1941)

[Hf]   Hopf, H.: Zur Topologie der komplexen Mannigfaltigkeiten. Studies and Essays Presented to R. Courant on his 60th Birthday, January 8, 1948, pp. 167–185. Interscience Publishers, New York (1948)

[Hö-1]   Hörmander, L.: L^2 estimates and existence theorems for the $\bar{\partial}$ operator. Acta Math. **113**, 89–152 (1965)

[Hö-2]   Hörmander, L.: An Introduction to Complex Analysis in Several Variables. North-Holland Mathematical Library, vol. 7, 3rd edn, xii+254 pp. North-Holland Publishing, Amsterdam (1990)

[Hö-3]   Hörmander, L.: A history of existence theorems for the Cauchy-Riemann complex in L^2 spaces. J. Geom. Anal. **13**, 329–357 (2003)

[H-W]   Hörmander, L., Wermer, J.: Uniform approximation on compact sets in $\mathbb{C}^n$. Math. Scand. **23**, 5–21 (1968)

[H-P]   Hsiang, W.C., Pati, V.: L^2-cohomology of normal algebraic surfaces. I. Invent. Math. **81**, 395–412 (1985)

[Hu]   Huber, A.: On subharmonic functions and differential geometry in the large. Comment. Math. Helv. **32**, 13–72 (1957)

[Huckl]   Huckleberry, A.: On certain domains in cycle spaces of flag manifolds. Math. Ann. **323**, 797–810 (2002)

[J-M]   Jonsson, M., Mustatá, M.: An algebraic approach to the openness conjecture of Demailly and Kollár. J. Inst. Math. Jussieu **13**, 119–144 (2014)

[K-Oh]   Kai, C., Ohsawa, T.: A note on the Bergman metric of bounded homogeneous domains. Nagoya Math. J. **186**, 157–163 (2007)

[Ka]   Kato, Ma.: Topology of Hopf surfaces. J. Math. Soc. Jpn. **27**, 222–238 (1975). Erratum, J. Math. Soc. Jpn. **41**, 173–174 (1989)

[Kt]   Kato, Mi.: Some problems in topology. In: Hattori, A. (ed.) Manifolds-Tokyo 1973,
            Proceedings of the International Conference on Manifolds and Related Topics in
            Topology Tokyo, 1973, pp. 421–431. University of Tokyo Press, Tokyo (1975)

[K-K]   Kawai, T., Kashiwara, M.: The Poincaré lemma for variations of polarized Hodge
            structure. Publ. Res. Inst. Math. Sci. **23**, 345–407 (1987)

[Kz-1]   Kazama, H.: Approximation theorem and application to Nakano's vanishing theorem
            for weakly 1-complete manifolds. Mem. Fac. Sci. Kyushu Univ. **27**, 221–240 (1973)

[Kz-2]   Kazama, H.: $\bar{\partial}$-cohomology of (H,C)-groups. Publ. Res. Inst. Math. Sci. **20**, 297–317
            (1984)

[Kz-3]   Kazama, H.: On pseudoconvexity of complex Lie groups. Mem. Fac. Sci. Kyushu
            Univ. **27**, 241–247 (1973)

[Km-1]   Kawamata, Y.: A generalization of Kodaira-Ramanujam's vanishing theorem. Math.
            Ann. **261**, 43–46 (1982)

[Km-2]   Kawamata, Y.: The cone of curves of algebraic varieties. Ann. Math. **119**, 603–633
            (1984)

[Km-3]   Kawamata, Y.: Pluricanonical systems on minimal algebraic varieties. Invent. Math.
            79, 567–588 (1985)

[Km-4]   Kawamata, Y.: On Fujita's freeness conjecture for 3-folds and 4-folds. Math. Ann.
            308, 491–505 (1997)

[Km-5]   Kawamata, Y.: Higher Dimensional Algebraic Varieties. (Japanese) Iwanami, Tokyo
            (2014)

[Km-M-M]   Kawamata, Y., Matsuda, K., Matsuki, K.: Introduction to the Minimal Model Prob-
            lem. Algebraic Geometry, Sendai, 1985. Advanced Studies in Pure Mathematics,
            vol. 10, pp. 283–360. North-Holland, Amsterdam (1987)

[Kp]   Kempf, G.R.: Complex Abelian Varieties and Theta Functions. Universitext, x+105
            pp. Springer, Berlin (1991)

[Kzm]   Kerzman, N.: The Bergman kernel function. Differentiability at the boundary. Math.
            Ann. **195**, 149–158 (1972)

[K-R]   Kerzman, N., Rosay, J.-P.: Fonctions plurisousharmoniques d'exhaustion bornées et
            domaines taut. Math. Ann. **257**, 171–184 (1981)

[K-L-Y]   Kim, K.T., Levenberg, N., Yamaguchi, H.: Robin Functions for Complex Manifolds
            and Applications. Memoirs of the American Mathematical Society, vol. 209, No. 984.
            American Mathematical Society, Providence (2011)

[Kb-1]   Kobayashi, S.: Geometry of bounded domains. Trans. Am. Math. Soc. **92**, 267–290
            (1959)

[Kb-2]   Kobayashi, S.: Differential Geometry of Complex Vector Bundles. Publications of
            the Mathematical Society of Japan, vol. 15, xii+305 pp. Kanô Memorial Lectures 5.
            Princeton University Press/Iwanami Shoten, Princeton/Tokyo (1987)

[K-1]   Kodaira, K.: Harmonic fields in Riemannian manifolds (generalized potential the-
            ory). Ann. Math. **50**, 587–665 (1949)

[K-2]   Kodaira, K.: On a differential-geometric method in the theory of analytic stacks.
            Proc. Natl. Acad. Sci. U.S.A. **39**, 1268–1273 (1953)

[K-3]   Kodaira, K.: On Kähler varieties of restricted type (an intrinsic characterization of
            algebraic varieties). Ann. Math. **60**, 28–48 (1954)

[K-4]   Kodaira, K.: On the structure of compact complex analytic surfaces, II, III. Am. J.
            Math. **88**, 682–721 (1966); **90**, 55–83 (1968)

[Kn]   Kohn, J.: Sufficient conditions for subellipticity on weakly pseudo-convex domains.
            Proc. Natl. Acad. Sci. U.S.A. **74**, 2214–2216 (1977)

[K-N]   Kohn, J.J., Nirenberg, L.: A pseudo-convex domain not admitting a holomorphic
            support function. Math. Ann. **201**, 265–268 (1973)

[Ko]   Kollár, J.: Effective base point freeness. Math. Ann. **296**, 595–605 (1993)

[Kr]   Krushkal', S.L.: Strengthening pseudoconvexity of finite-dimensional Teichmüller
            spaces. Math. Ann. **290**, 681–687 (1991)

[L]   Lattes, M.S.: Sur les formes réduits des transformations ponctuelles à deux variables. C. R. Acad. Sci. **152**, 1566–1569 (1911)

[LT-L] Laurent-Thiébaut, C., Leiterer, J.: Some applications of Serre duality in CR manifolds. Nagoya Math. J. **154**, 141–156 (1999)

[L-Y-1] Levenberg, N., Yamaguchi, H.: The Metric Induced by the Robin Function. Memoirs of the American Mathematical Society, vol. 92, 448, viii+156 pp. American Mathematical Society, Providence (1991)

[L-Y-2] Levenberg, N., Yamaguchi, H.: Pseudoconvex domains in the Hopf surface. J. Math. Soc. Jpn. **67**, 231–273 (2015)

[L-M]   Lieb, I., Michel, J.: The Cauchy-Riemann Complex. Integral Formulae and Neumann Problem. Aspects of Mathematics, E34, x+362 pp. Friedr. Vieweg & Sohn, Braunschweig (2002)

[LN]   Lins Neto, A.: A note on projective Levi flats and minimal sets of algebraic foliations. Ann. Inst. Fourier **49**, 1369–1385 (1999)

[L-T]   Lipman, J., Tessier, B.: Pseudo-rational local rings and a theorem of Briançon-Skoda. Mich. Math. J. **28**, 97–115 (1981)

[L-S-Y] Liu, K., Sun, X., Yang, X.: Positivity and vanishing theorems for ample vector bundles. J. Algebr. Geom. **22**, 303–331 (2013)

[L-Y]   Liu, K., Yang, X.: Curvatures of direct image sheaves of vector bundles and applications. J. Differ. Geom. **98**, 117–145 (2014)

[Lj]   Looijenga, E.: L^2-cohomology of locally symmetric varieties. Compositio Math. **67**, 3–20 (1988). About integral closures of ideals, Mich. Math. J. **28**, 97–115 (1981)

[Ma]   Mabuchi, T.: Asymptotic structures in the geometry of stability and extremal metrics (English summary). In: Handbook of Geometric Analysis. Advanced Lectures in Mathematics (ALM), vol. 7, No. 1, pp. 277–300. International Press, Somerville (2008)

[M-Y]   Maitani, F., Yamaguchi, H.: Variation of Bergman metrics on Riemann surfaces. Math. Ann. **330**, 477–489 (2004)

[M-M]   Matsushima, Y., Morimoto, A.: Sur certains espaces fibrés holomorphes sur une variété de Stein. Bull. Soc. Math. Fr. **88**, 137–155 (1960)

[M]   Matsumoto, K.: Levi form of logarithmic distance to complex submanifolds and its application to developability. In: Complex Analysis in Several Variables – Memorial Conference of Kiyoshi Oka's Centennial Birthday. Advanced Studies in Pure Mathematics, vol. 42, pp. 203–207. Mathematical Society of Japan, Tokyo (2004)

[M-P]   Merker, J., Porten, E.: The Hartogs extension theorem on (n-1)-complete complex spaces. J. Reine Angew. Math. **637**, 23–39 (2009)

[Mb]   Miebach, C.: Pseudoconvex non-Stein domains in primary Hopf surfaces. Izv. Math. **78**, 1028–1035 (2014)

[Ml]   Milnor, J.: On deciding whether a surface is parabolic or hyperbolic. Am. Math. Mon. **84**, 43–46 (1977)

[Mk]   Mok, N.: The Serre problem on Riemann surfaces. Math. Ann. **258**, 145–168 (1981/1982)

[M-Y]   Mok, N., Yau, S.-T.: Completeness of the Kähler-Einstein metric on bounded domains and the characterization of domains of holomorphy by curvature conditions. In: The Mathematical Heritage of Henri Poincaré, Part 1, Bloomington, 1980. Proceedings of Symposia in Pure Mathematics, vol. 39, pp. 41–59. American Mathematical Society, Providence (1983)

[M-1]   Mori, S.: Projective manifolds with ample tangent bundles. Ann. Math. **110**, 593–606 (1979)

[M-2]   Mori, S.: Threefolds whose canonical bundles are not numerically effective. Ann. Math. **116**, 133–176 (1982)

[Mr]   Morimoto, A.: Non-compact complex Lie groups without non-constant holomorphic
          functions. In: Proceedings of the Conference on Complex Analysis, Minneapolis
          1964, pp. 256–272. Springer (1965)

[Mry]   Morrey, C.B., Jr.: The analytic embedding of abstract real-analytic manifolds. Ann.
          Math. **68**, 159–201 (1958)

[M-T]   Mourougane, C., Takayama, S.: Hodge metrics and positivity of direct images. J. für
          die reine u. angew. Mathematik **606**, 167–178 (2007)

[Mm]   Mumford, D.: The topology of normal singularities of an algebraic surface and a
          criterion for simplicity. Inst. Hautes Études Sci. Publ. Math. **9**, 5–22 (1961)

[Nd]   Nadel, A.: Multiplier ideal sheaves and Kähler-Einstein metrics of positive scalar
          curvature. Ann. Math. **132**, 549–596 (1990)

[Ns]   Nagase, M.: Remarks on the L^2-cohomology of singular algebraic surfaces. J. Math.
          Soc. Jpn. **41**, 97–116 (1989)

[Ng]   Nagata, M.: Local Rings. Interscience Tracts in Pure and Applied Mathematics,
          No. 13, xiii+234 pp. Interscience Publishers a division of John Wiley & Sons, New
          York/London (1962)

[Na-1]   Nakai, Y.: Non-degenerate divisors on an algebraic surface. J. Sci. Hiroshima Univ.
          24, 1–6 (1960)

[Na-2]   Nakai, Y.: A criterion of an ample sheaf on a projective scheme. Am. J. Math. **85**,
          14–26 (1963)

[Na-3]   Nakai, Y.: Some fundamental lemmas on projective schemes. Trans. Am. Math. Soc.
          109, 296–302 (1963)

[Nk]   Nakamura, I.: Moduli and Deformation Theory (Japanese), Have Fun with Mathe-
          matics 20, pp. 38–53. Nippon Hyoron Sha Co. Tokyo (2000)

[N-1]   Nakano, S.: On complex analytic vector bundles. J. Math. Soc. Jpn. **7**, 1–12 (1955)

[N-2]   Nakano, S.: On the inverse of monoidal transformation. Publ. Res. Inst. Math. Sci. **6**,
          483–502 (1970/1971)

[N-3]   Nakano, S.: Vanishing theorems for weakly 1-complete manifolds. II. Publ. Res. Inst.
          Math. Sci. **10**, 101–110 (1974/1975)

[N-Oh]   Nakano, S., Ohsawa, T.: Strongly pseudoconvex manifolds and strongly pseudocon-
          vex domains. Publ. Res. Inst. Math. Sci. **20**, 705–715 (1984)

[N-R]   Nakano, S., Rhai, T.-S.: Vector bundle version of Ohsawa's finiteness theorems.
          Math. Jpn. **24**, 657–664 (1979/1980)

[N]   Narasimhan, R.: Sur les espaces complexes holomorphiquement complets. C. R.
          Acad. Sci. Paris **250**, 3560–3561 (1960)

[Nm]   Nemirovski, S.: Stein domains with Levi-plane boundaries on compact complex
          surfaces. Mat. Zametki **66**, 632–635 (1999); translation in Math. Notes **66**(3–4), 522–
          525 (1999/2000)

[Ni-1]   Nishino, T.: Sur une propriété des familles de fonctions analytiques de deux variables
          complexes. J. Math. Kyoto Univ. **4**, 255–282 (1965)

[Ni-2]   Nishino, T.: L'existence d'une fonction analytique sur une variété analytique com-
          plexe à deux dimensions. Publ. Res. Inst. Math. Sci. **18**, 387–419 (1982)

[Ni-3]   Nishino, T.: L'existence d'une fonction analytique sur une variété analytique com-
          plexe à dimension quelconque. Publ. Res. Inst. Math. Sci. **19**, 263–273 (1983)

[Nog]   Noguchi, J.: Analytic function theory of several variables —Elements of Oka's
          coherence, preprint

[Oh-0]   Ohsawa, T.: A counter example of ampleness of positive line bundles. Proc. Jpn.
          Acad. Ser. A Math. Sci. **55**, 193–194 (1979)

[Oh-1]   Ohsawa, T.: Finiteness theorems on weakly 1-complete manifolds. Publ. Res. Inst.
          Math. Sci. **15**, 853–870 (1979)

[Oh-2]   Ohsawa, T.: On complete Kähler domains with C^1-boundary. Publ. Res. Inst. Math.
          Sci. **16**, 929–940 (1980)

[Oh-3]   Ohsawa, T.: Analyticity of complements of complete Kähler domains. Proc. Jpn.
          Acad. Ser. A Math. Sci. **56**, 484–487 (1980)

[Oh-4] Ohsawa, T.: On $H^{p,q}(X, B)$ of weakly 1-complete manifolds. Publ. Res. Inst. Math. Sci. **17**, 113–126 (1981)

[Oh-5] Ohsawa, T.: A reduction theorem for cohomology groups of very strongly q-convex Kähler manifolds. Invent. Math. **63**, 335–354 (1981)

[Oh-6] Ohsawa, T.: Addendum to: "A reduction theorem for cohomology groups of very strongly q-convex Kähler manifolds" (Invent. Math. **63**(2), 335–354 (1981)). Invent. Math. **66**, 391–393 (1982)

[Oh-7] Ohsawa, T.: Isomorphism theorems for cohomology groups of weakly 1-complete manifolds. Publ. Res. Inst. Math. Sci. **18**, 191–232 (1982)

[Oh-8] Ohsawa, T.: Vanishing theorems on complete Kähler manifolds. Publ. Res. Inst. Math. Sci. **20**, 21–38 (1984)

[Oh-9] Ohsawa, T.: Completeness of noncompact analytic spaces. Publ. Res. Inst. Math. Sci. **20**, 683–692 (1984)

[Oh-10] Ohsawa, T.: Hodge spectral sequence on compact Kähler spaces. Publ. Res. Inst. Math. Sci. **23**, 265–274 (1987). *Supplement*: Publ. Res. Inst. Math. Sci. **27**, 505–507 (1991)

[Oh-11] Ohsawa, T.: Hodge spectral sequence and symmetry on compact Kähler spaces. Publ. Res. Inst. Math. Sci. **23**, 613–625 (1987)

[Oh-12] Ohsawa, T.: On the infinite dimensionality of the middle L^2 cohomology of complex domains. Publ. Res. Inst. Math. Sci. **25**, 499–502 (1989)

[Oh-13] Ohsawa, T.: Intersection cohomology – where L^2 theory meets the mixed Hodge theory (Japanese). In: Research on Complex Analytic Geometry and Related Topics, RIMS Kokyuroku, vol. 693, pp. 23–40. Research Institute for Mathematical Sciences, Kyoto (1989)

[Oh-14] Ohsawa, T.: Cheeger-Goreski [Goresky]-MacPherson's conjecture for the varieties with isolated singularities. Math. Z. **206**, 219–224 (1991)

[Oh-15] Ohsawa, T.: On the L^2 cohomology groups of isolated singularities. In: Progress in Differential Geometry. Advanced Studies in Pure Mathematics, vol. 22, pp. 247–263. Mathematical Society of Japan, Tokyo (1993)

[Oh-16] Ohsawa, T.: On the Bergman kernel of hyperconvex domains. Nagoya Math. J. **129**, 43–52 (1993). Addendum, Nagoya Math. J. **137**, 145–148 (1995)

[Oh-17] Ohsawa, T.: On the extension of L^2 holomorphic functions III. Negligible weights. Math. Z. **219**, 215–225 (1995)

[Oh-18] Ohsawa, T.: Pseudoconvex domains in $\mathbb{P}^n$: a question on the 1-convex boundary points. In: Analysis and Geometry in Several Complex Variables, Katata, 1997. Trends in Mathematics, pp. 239–252. Birkhäuser, Boston (1999)

[Oh-19] Ohsawa, T.: On the extension of L^2 holomorphic functions V. Effect of generalization. Nagoya Math. J. **161**, 1–21 (2001)

[Oh-20] Ohsawa, T.: A precise L^2 division theorem. In: Complex Geometry, Göttingen, 2000, pp. 185–191. Springer, Berlin (2002)

[Oh-21] Ohsawa, T.: Analysis of Several Complex Variables (Translated from the Japanese by Shu Gilbert Nakamura). Translations of Mathematical Monographs, vol. 211. Iwanami Series in Modern Mathematics, xviii+121 pp. American Mathematical Society, Providence (2002)

[Oh-22] Ohsawa, T.: On the complement of Levi-flats in Kähler manifolds of dimension ≥ 3. Nagoya Math. J. **185**, 161–169 (2007)

[Oh-23] Ohsawa, T.: A remark on pseudoconvex domains with analytic complements in compact Kähler manifolds. J. Math. Kyoto Univ. **47**, 115–119 (2007)

[Oh-24] Ohsawa, T.: A reduction theorem for stable sets of holomorphic foliations on complex tori. Nagoya Math. J. **195**, 41–56 (2009)

[Oh-25] Ohsawa, T.: A tower of Riemann surfaces whose Bergman kernels jump at the roof. Publ. Res. Inst. Math. Sci. **46**, 473–478 (2010)

[Oh-26] Ohsawa, T.: Hartogs type extension theorems on some domains in Kähler manifolds. Ann. Polon. Math. **106**, 243–254 (2012)

[Oh-27]   Ohsawa, T.: On the complement of effective divisors with semipositive normal bundle. Kyoto J. Math. **52**, 503–515 (2012)

[Oh-28]   Ohsawa, T.: On projectively embeddable complex-foliated structures. Publ. Res. Inst. Math. Sci. **48**, 735–747 (2012)

[Oh-29]   Ohsawa, T.: Nonexistence of certain Levi flat hypersurfaces in Kähler manifolds from the viewpoint of positive normal bundles. Publ. Res. Inst. Math. Sci. **49**, 229–239 (2013)

[Oh-30]   Ohsawa, T.: Classification of real analytic Levi flat hypersurfaces of 1-concave type in Hopf surfaces. Kyoto J. Math. **54**, 547–553 (2014)

[Oh-31]   Ohsawa, T.: A lemma on Hartogs function and application to Levi flat hypersurfaces in Hopf surfaces, to appear in Prog. Math. **38** (2015)

[Oh-32]   Ohsawa, T.: Stability of pseudoconvexity of disc bundles over compact Riemann surfaces and application to a family of Galois coverings. Int. J. Math. **26**, 1540003 (2015), [7 pages] doi:10.1142/S0129167X15400030

[Oh-33]   Ohsawa, T.: Application and simplified proof of a sharp L^2 extension theorem, to appear in Nagoya Math. J.

[Oh-34]   Ohsawa, T.: A remark on Hörmander's isomorphism, to appear in proceedings of KSCV 10

[Oh-35]   Ohsawa, T.: Some updates of extension theorems by the L^2 estimates for $\bar{\partial}$, submitted for publication

[Oh-S]   Ohsawa, T., Sibony, N.: Bounded p.s.h. functions and pseudoconvexity in Kähler manifold. Nagoya Math. J. **149**, 1–8 (1998)

[Oh-T-1]   Ohsawa, T., Takegoshi, K.: On the extension of L^2 holomorphic functions. Math. Z. **195**, 197–204 (1987)

[Oh-T-2]   Ohsawa, T., Takegoshi, K.: Hodge spectral sequence on pseudoconvex domains. Math. Z. **197**, 1–12 (1988)

[O-S]   Oikawa, K., Sario, L.: Capacity Functions. Die Grundlehren der mathematischen Wissenschaften, Band 149, xvii+361 pp. Springer, New York (1969)

[O-1]   Oka, K.: Sur les fonctions analytiques de plusieurs variables VI. Domaines pseudo-convexes. Tôhoku Math. J. **49**, 15–52 (1942)

[O-2]   Oka, K.: Sur les fonctions analytiques de plusieurs variables VII. Sur quelques notions arithmétiques. Bull. Soc. Math. Fr. **78**, 1–27 (1950)

[O-3]   Oka, K.: Sur les fonctions analytiques de plusieurs variables VIII. Lemme fondamental. J. Math. Soc. Jpn. **3**, 204–214 and 259–278 (1951)

[O-4]   Oka, K.: Sur les fonctions analytiques de plusieurs variables IX. Domaines finis sans point critique intérieur. Jpn. J. Math. **23**, 97–155 (1953)

[P]   Păun, M.: Siu's invariance of plurigenera: a one-tower proof. J. Differ. Geom. **76**, 485–493 (2007)

[Pf]   Pflug, P.: Quadratintegrable holomorphe Funktionen und die Serre-Vermutung. Math. Ann. **216**, 285–288 (1975)

[PS]   Pjateckii-Shapiro, I.I.: Géométrie des domaines classiques et théorie des fonctions automorphes, Traduit du Russe par A. W. Golovanoff. Travaux et Recherches Mathêmatiques, No. 12, iv+160 pp. Dunod, Paris (1966)

[Rm-1]   Ramanujam, C.P.: Remarks on the Kodaira vanishing theorem. J. Indian Math. Soc. **36**, 41–51 (1972)

[Rm-2]   Ramanujam, C.P.: Supplement to the article "Remarks on the Kodaira vanishing theorem". J. Indian Math. Soc. **38**, 121–124 (1974)

[Rn]   Raynaud, M.: Contre-exemple au "vanishing theorem" en caractéristique $p > 0$, C. P. Ramanujam – a tribute, 273–278, Tata Inst. Fund. Res. Studies in Math., vol. 8. Springer, Berlin/New York (1978)

[Rd]   Reider, I.: Vector bundles of rank 2 and linear systems on algebraic surfaces. Ann. Math. **127**, 309–316 (1988)

[R-1]   Remmert, R.: Sur les espaces analytiques holomorphiquement séparables et holomorphiquement convexes. C. R. Acad. Sci. Paris **243**, 118–121 (1956)

[R-2]   Remmert, R.: From Riemann surfaces to complex spaces. In: Matériaux pour l'histoire des mathématiques au XXe siècle, Nice, 1996. Sémin. Congr., vol. 3, pp. 203–241. Société Mathématique de France, Paris (1998)

[R-S]   Remmert, R., Stein, K.: Über die wesentlichen Singularitäten analytischer Mengen. Math. Ann. **126**, 263–306 (1953)

[Rh]   Rhodes, J.A.: Sequences of metrics on compact Riemann surfaces. Duke Math. J. **72**, 725–738 (1993)

[R]   Richberg, R.: Stetige streng pseudokonvexe Funktionen. Math. Ann. **175**, 257–286 (1968)

[Rt]   Rothstein, W.: Zur Theorie der Analytischen Mannigfaltigkeiten im Raume von n komplexen Veränderlichen. Math. Ann. **129**, 96–138 (1955)

[Sai]   Saito, M.: Mixed Hodge modules. Publ. Res. Inst. Math. Sci. **26**, 221–333 (1990)

[Sak]   Sakai, A.: Uniform approximation on totally real sets. Math. Ann. **253**, 139–144 (1980)

[Sap]   Saper, L.: L^2-cohomology and intersection homology of certain algebraic varieties with isolated singularities. Invent. Math. **82**, 207–256 (1985)

[Sap-St] Saper, L., Stern, M.: L^2-cohomology of arithmetic varieties. Proc. Natl. Acad. Sci. U.S.A. **84**, 5516–5519 (1987)

[S-O]   Sario, L., Oikawa, K.: Capacity Functions. Die Grundlehren der mathematischen Wissenschaften, Band 149, xvii+361 pp. Springer, New York (1969)

[Sat]   Satake, I.: Compactifications, old and new (translation of Sūgaku **51**, 129–141). Sugaku Expositions **14**, 175–189 (2001)

[Sch]   Scherk, J.: On the monodromy theorem for isolated hypersurface singularities, Invent. Math. **58**, 289–301 (1980)

[S]   Schoutens, H.: A non-standard proof of the Briançon-Skoda theorem. Proc. Am. Math. Soc. **131**, 103–112 (2003)

[Sm]   Schürmann, J.: Embeddings of Stein spaces into affine spaces of minimal dimension. Math. Ann. **307**, 381–399 (1997)

[Sp-1]   Seip, K.: Density theorems for sampling and interpolation in the Bargmann-Fock space I. J. Reine Angew. Math. **429**, 91–106 (1992)

[Sp-2]   Seip, K.: Beurling type density theorems in the unit disk. Invent. Math. **113**, 21–39 (1993)

[S-W]   Seip, K., Wallstén, R.: Density theorems for sampling and interpolation in the Bargmann-Fock space II. J. Reine Angew. Math. **429**, 107–113 (1992)

[S-1]   Serre, J.-P.: Quelques problèmes globaux relatifs aux variétés de Stein, Colloque sur les fonctions de plusieurs variables, Bruxelles, pp. 53–68 (1953)

[S-2]   Serre, J.-P.: Un théorème de dualité. Comment. Math. Helv. **29**, 9–26 (1955)

[Sha]   Shafarevich, I.R.: Basic Algebraic Geometry (Translated from the Russian by K. A. Hirsch). Die Grundlehren der mathematischen Wissenschaften, Band 213, xv+439 pp. Springer, New York/Heidelberg (1974)

[Shc]   Shcherbina, N.: Pluripolar graphs are holomorphic. Acta Math. **194**, 203–216 (2005)

[Siu-1] Siu, Y.-T.: Analytic sheaf cohomology groups of dimension n of n-dimensional noncompact complex manifolds. Pac. J. Math. **28**, 407–411 (1969)

[Siu-2] Siu, Y.-T.: Techniques of Extension of Analytic Objects. Lecture Notes in Pure and Applied Mathematics, vol. 8, iv+256 pp. Marcel Dekker, New York (1974)

[Siu-3] Siu, Y.-T.: Characterization of privileged polydomains. Trans. Am. Math. Soc. **193**, 329–357 (1974)

[Siu-4] Siu, Y.-T.: Analyticity of sets associated to Lelong numbers and the extension of closed positive currents. Invent. Math. **27**, 53–156 (1974)

[Siu-5] Siu, Y.-T.: Complex-analyticity of harmonic maps, vanishing and Lefschetz theorems. J. Differ. Geom. **17**, 55–138 (1982)

[Siu-6] Siu, Y.-T.: The Fujita conjecture and the extension theorem of Ohsawa-Takegoshi. In: Geometric Complex Analysis, Hayama, 1995, pp. 577–592. World Scientific Publishing, River Edge (1996)

[Siu-7]   Siu, Y.-T.: Invariance of plurigenera. Invent. Math. **134**, 661–673 (1998)
[Siu-8]   Siu, Y.-T.: Nonexistence of smooth Levi-flat hypersurfaces in complex projective
           spaces of dimension $\geq$ 3. Ann. Math. **151**, 1217–1243 (2000)
[Siu-9]   Siu, Y.-T.: Extension of twisted pluricanonical sections with plurisubharmonic
           weight and invariance of semipositively twisted plurigenera for manifolds not
           necessarily of general type. In: Complex Geometry, Göttingen, 2000, pp. 223–277.
           Springer, Berlin (2002)
[Siu-10]   Siu, Y.-T.: Some recent transcendental techniques in algebraic and complex geome-
           try. In: Proceedings of the International Congress of Mathematicians, Beijing, 2002,
           vol. I, pp. 439–448. Higher Education Press, Beijing (2002)
[Siu-11]   Siu, Y.-T.: Multiplier ideal sheaves in complex and algebraic geometry. Sci. China
           Ser. A **48**(suppl.), 1–31 (2005)
[Siu-12]   Siu, Y.-T.: Finite generation of canonical ring by analytic method. Sci. China Ser. A
           51, 481–502 (2008)
[S-Y]   Siu, Y.T., Yau, S.T.: Compact Kähler manifolds of positive bisectional curvature.
           Invent. Math. **59**, 189–204 (1980)
[St]   Stout, E.L.: Polynomial Convexity. Progress in Mathematics, vol. 261, xii+439 pp.
           Birkhäuser, Boston (2007)
[Sk-1]   Skoda, H.: Sous-ensembles analytiques d'ordre fini ou infini dans $\mathbb{C}^n$. Bull. Soc.
           Math. Fr. **100**, 353–408 (1972)
[Sk-2]   Skoda, H.: Application des techniques L^2 à la théorie des idéaux d'une algèbre de
           fonctions holomorphes avec poids. Ann. Sci. École Norm. Sup. **5**, 545–579 (1972)
[Sk-3]   Skoda, H.: Fibrés holomorphes à base et à fibre de Stein. Invent. Math. **43**, 97–107
           (1977)
[Sk-4]   Skoda, H.: Morphismes surjectifs de fibrés vectoriels semi-positifs. Ann. Sci. École
           Norm. Sup. **11**, 577–611 (1978)
[Sk-5]   Skoda, H.: Problem Session of the Conference on Complex Analysis, pp. 1–3 (2005).
           www.institut.math.jussieu.fr/projets/...skoda/probleme.Skoda
[Su-1]   Suita, N.: Capacities and kernels on Riemann surfaces. Arch. Ration. Mech. Anal.
           46, 212–217 (1972)
[Su-2]   Suita, N.: Modern Function Theory II, (POD edition). Morikita Publishing (2011)
           (Japanese)
[Suz]   Suzuki, O.: Pseudoconvex domains on a Kähler manifold with positive holomorphic
           bisectional curvature. Publ. Res. Inst. Math. Sci. **12**, 191–214 (1976/1977). Supple-
           ment: Publ. Res. Inst. Math. Sci. **12**, 439–445 (1976/1977)
[Sz]   Sznajdman, J.: An elementary proof of the Briançon-Skoda theorem. Annales de la
           facult é des sciences de Toulouse Sér. 6 **19**, 675–685 (2010)
[Ty-1]   Takayama, S.: Adjoint linear series on weakly 1-complete Kähler manifolds. I.
           Global projective embedding. Math. Ann. **311**, 501–531 (1998)
[Ty-2]   Takayama, S.: Adjoint linear series on weakly 1-complete Kähler manifolds. II.
           Lefschetz type theorem on quasi-abelian varieties. Math. Ann. **312**, 363–385 (1998)
[Ty-3]   Takayama, S.: The Levi problem and the structure theorem for non-negatively curved
           complete Kähler manifolds. J. Reine Angew. Math. **504**, 139–157 (1998)
[T-1]   Takegoshi, K.: A generalization of vanishing theorems for weakly 1-complete
           manifolds. Publ. Res. Inst. Math. Sci. **17**, 311–330 (1981)
[T-2]   Takegoshi, K.: Relative vanishing theorems in analytic spaces. Duke Math. J. **52**,
           273–279 (1985)
[Tk-1]   Takeuchi, A.: Domaines pseudoconvexes infinis et la métrique riemannienne dans un
           espace projectif. J. Math. Soc. Jpn. **16**, 159–181 (1964)
[Tk-2]   Takeuchi, A.: Domaines pseudoconvexes sur les variétés kählériennes. J. Math.
           Kyoto Univ. **6**, 323–357 (1967)
[Ti]   Tian, G.: On a set of polarized Kähler metrics on algebraic manifolds. J. Differ.
           Geom. **32**, 99–130 (1990)

[U-1]   Ueda, T.: On the neighborhood of a compact complex curve with topologically trivial normal bundle. J. Math. Kyoto Univ. **22**, 583–607 (1982/1983)

[U-2]   Ueda, T.: Pseudoconvex domains over Grassmann manifolds. J. Math. Kyoto Univ. **20**, 391–394 (1980)

[Va]   Varouchas, J.: Stabilité de la classe des variétés kählériennes par certains morphismes propres. Invent. Math. **77**, 117–127 (1984)

[V]   Viehweg, E.: Vanishing theorems. J. Reine Angew. Math. **335**, 1–8 (1982)

[Vo]   Vogt, C.: Two renarks concerning toroidal groups. Manuscr. Math. **41**, 217–232 (1983)

[Wl]   Wall, C.T.C.: Lectures on C^∞-stability and classification. In: Proceedings of Liverpool Singularities Symposium I (1969/1970). Lecture Notes in Mathematics, vol. 192, pp. 178–206. Springer, Berlin (1971)

[W-1]   Weil, A.: Sur la théorie des formes différentielles attachées à une variété analytique complexe. Comment. Math. Helv. **20**, 110–116 (1947)

[W-2]   Weil, A.: Introduction à l'étude des variétés kählériennes, Publications de l'Institut de Mathématique de l'Université de Nancago, VI. Actualités Sci. Ind. no. 1267, 175 p. Hermann, Paris (1958)

[W]   Wells, R.O., Jr.: Differential Analysis on Complex Manifolds, 3rd edn. With a new appendix by Oscar Garcia-Prada. Graduate Texts in Mathematics, vol. 65. Springer, New York (2008)

[Wy-1]   Weyl, H.: The method of orthogonal projection in potential theory. Duke Math. J. **7**, 411–444 (1940)

[Wy-2]   Weyl, H.: David Hilbert and his mathematical work. Bull. Am. Math. Soc. **50**, 612–654 (1944)

[Wn]   Wiener, N.: Certain notions in potential theory. J. Math. MIT **3**, 24–51 (1924)

[Wu]   Wu, H.: An elementary method in the study of nonnegative curvature. Acta Math. **142**, 57–78 (1979)

[Y-1]   Yamaguchi, H.: Sur le mouvement des constantes de Robin. J. Math. Kyoto Univ. **15**, 53–71 (1975)

[Y-2]   Yamaguchi, H.: Parabolicité d'une fonction entière. J. Math. Kyoto Univ. **16**, 71–92 (1976)

[Y-3]   Yamaguchi, H.: Variations of pseudoconvex domains over $\mathbb{C}^n$. Mich. Math. J. **36**, 415–457 (1989)

[Yau-1]   Yau, S.-T.: A general Schwarz lemma for Kähler manifolds. Am. J. Math. **100**, 197–203 (1978)

[Yau-2]   Yau, S.-T.: Nonlinear analysis in geometry. l'Enseignement Math. **33**, 109–158 (1986)

[Za]   Zarankiewicz, K.: Über ein numerishces Verfahren zur konformen Abbildung zweifach zusammenhängender Gebiete. Z. Angew. Math. Mech. **14**, 97–104 (1934)

[Ze]   Zelditch, S.: Szegö kernels and a theorem of Tian. Int. Math. Res. Not. **6**, 317–331 (1998)

[Z]   Zucker, S.: Hodge theory with degenerating coefficients, L^2-cohomology in the Poincaré metric. Ann. Math. **109**, 415–476 (1979)

Index

© Springer Japan 2015
T. Ohsawa, L^2 *Approaches in Several Complex Variables*, Springer Monographs
in Mathematics, DOI 10.1007/978-4-431-55747-0

Printed in the United States
By Bookmasters